A Guide to Innovation Processes and Solutions for Government

Dr. Gregory C. McLaughlin
Dr. William R. Kennedy

CRC Press
Taylor & Francis Group
Boca Raton London New York

CRC Press is an imprint of the
Taylor & Francis Group, an **informa** business

A PRODUCTIVITY PRESS BOOK

CRC Press
Taylor & Francis Group
6000 Broken Sound Parkway NW, Suite 300
Boca Raton, FL 33487-2742

© 2016 by Taylor & Francis Group, LLC
CRC Press is an imprint of Taylor & Francis Group, an Informa business

No claim to original U.S. Government works

Printed on acid-free paper
Version Date: 20150720

International Standard Book Number-13: 978-1-4987-2157-8 (Hardback)

This book contains information obtained from authentic and highly regarded sources. Reasonable efforts have been made to publish reliable data and information, but the author and publisher cannot assume responsibility for the validity of all materials or the consequences of their use. The authors and publishers have attempted to trace the copyright holders of all material reproduced in this publication and apologize to copyright holders if permission to publish in this form has not been obtained. If any copyright material has not been acknowledged please write and let us know so we may rectify in any future reprint.

Except as permitted under U.S. Copyright Law, no part of this book may be reprinted, reproduced, transmitted, or utilized in any form by any electronic, mechanical, or other means, now known or hereafter invented, including photocopying, microfilming, and recording, or in any information storage or retrieval system, without written permission from the publishers.

For permission to photocopy or use material electronically from this work, please access www.copyright.com (http://www.copyright.com/) or contact the Copyright Clearance Center, Inc. (CCC), 222 Rosewood Drive, Danvers, MA 01923, 978-750-8400. CCC is a not-for-profit organization that provides licenses and registration for a variety of users. For organizations that have been granted a photocopy license by the CCC, a separate system of payment has been arranged.

Trademark Notice: Product or corporate names may be trademarks or registered trademarks, and are used only for identification and explanation without intent to infringe.

Library of Congress Cataloging-in-Publication Data

McLaughlin, Gregory C.
 A guide to innovation processes and solutions for government / Greg C. McLaughlin and William R. Kennedy.
     pages cm
 Includes bibliographical references and index.
 ISBN 978-1-4987-2157-8
 1. Public contracts--United States--Management. 2. Subcontracting--United States--Management. 3. Contracting out--United States. 4. Organizational effectiveness--United States. 5. Project management--United States. I. Kennedy, William R., 1960- II. Title.

HD3861.U6M4 2016
658.4'063--dc23   2015010781

Visit the Taylor & Francis Web site at
http://www.taylorandfrancis.com

and the CRC Press Web site at
http://www.crcpress.com

Printed and bound in Great Britain by
TJ International Ltd, Padstow, Cornwall

Contents

Acknowledgments xi
About the Authors xiii

1 Why This Book Is Necessary 1
   Introduction 1
   Why Innovation Is Such a Difficult Goal to Achieve 4
   Ongoing Research 5
   Culture and the Environment 5
   Dynamic Capabilities 6
   Requirements 7
   Assessment 9
   Strategies for Innovation 10
   Innovation Process Management 11
   Incremental Innovation 12
   Summary 13

2 Innovation: A Combination of Perspectives 15
   Introduction 15
   Coalescence/Combination 16
      Opinions: Expanding the Definition of Innovation 17
   A Research Study 20
      Background 21
      Summarizing the Results 23
   The Contribution of Satisfaction 25
      Needs (Requirements) Satisfaction 25
   User (Customer) Satisfaction 27

The Amalgamation of Choices 30
The Importance of Information 31
The Influence of Judgment 33
    Knowledge 33
    Experiences 34
    Performance 36
Inclusion 38
Summary 38
Discussion Questions 39
Assignments 39

3 Cultural and Environmental Influences on Innovation 41
Introduction 41
Large-Scale Cultural Influences: Individual Perspective 43
    National Cultures 43
    Ethnicities (Diversity?) 44
    Organizational Cultures 46
    Type of Government (Political System) Influences 47
    Affiliate Cultures 49
    Family 51
    Peers (Friends) 51
Estimating the Effect of Culture on Innovation 52
    Culture and the Environment 53
    Use of Alliances in the Public Domain 55
A Leader's Role in Creating a Culture of Innovation 57
Summary 59
Discussion Questions 60
Assignments 61

4 Dynamic Capabilities: The Power of Innovation 63
Introduction 63
Dynamic Capabilities 65
Dynamic Capabilities and the Leader 66
A Case for Process Mapping 68

Benefits of Process Mapping for Innovation......................70
Lean Concepts..72
Cause and Effect 74
Summary 78
Discussion Questions..79
Assignments 80

5 Needs and Requirements: How to Initiate Sustained Successful Innovation 81
Introduction..81
Requirements, Core Competency, and Mission Set Definition 82
Acquiring Capabilities and Addressing Capability Gaps 85
The Defense Acquisition System 86
Joint Capabilities and Integration Development System... 88
Planning, Programming, Budget, and Execution Process..89
Managing the Complexities of the DAS, JCIDS, and PPBE Relationship 90
Sequestration..91
JCIDS and the Requirements Generation Process 93
Systemic Leadership Challenges within the Requirements Definition Process 98
Summary 99
Discussion Questions... 101
Assignment... 102

6 Organizational Assessment: Is the Organization Ready for Innovation?..103
Introduction.. 103
Organizational Evaluation...104
  Initial Assessment: Organizational Readiness for Innovation 104
  Interpretation..106
  Actions...107

Evaluating Innovation Success 108
　Interpretation 110
　Actions 111
Evaluating Agency and Organizational Culture 112
　Assessing the Work Environment 112
　Interpretation 114
　Actions 114
Innovation Readiness 114
　Step 1: Needs (Requirements) and New Ideas 115
　Step 2: Nominate and Normalize 115
　Step 3: Objectify and Operationalize 116
　Step 4: Verify and Validate 116
　Step 5: Align and Adapt 117
　Step 6: Track and Transfer Performance 117
　Step 7: Evaluate and Execute 118
　Scoring 118
　Interpretation 119
　Actions 122
N²OVATE™ IR Proficiency 122
Assessment of Needs (Requirements) and Values 124
Summary 126
Discussion Questions 127
Assignments 127

7 Organizational Diagnostics: Through the Looking Glass 129
　Introduction 129
　Why Perform Diagnostics? 131
　Diagnostic Elements 132
　　Situational Analysis 132
　　Environmental Scan 133
　　Active Data (Information) Sources 134
　　Passive Data (Information) Sources 135
　　Summarize the Information 135
　　Assessment 136
　　Building a Profile 136

    Causal Inquiry 138
    Summary 139
    Discussion Questions 139
    Assignments 139

8 Innovation Strategies: Design for Success 141
    Introduction.. 141
    Organizational Scale 142
        Existing Strategies 143
    Innovation Strategy Types 145
        Strategy 1: Ad Hoc 147
        Strategy 2: Supplier Based 147
        Strategy 3: Directive Driven 147
        Strategy 4: R&D Emphasis 148
        Strategy 5: Science Based 148
        Strategy 6: No Strategy 148
    Mission, Vision, and Purpose Statements 150
    Compatibility with Organizational Outcomes and
    Functions 153
    Constructing an Innovation Strategy 155
        Strengths, Weaknesses, Opportunities, and Threats
        Analysis 156
        Assembling the Strategy 158
        Persistence 159
    Implementing an Innovation Strategy 159
    Success Factors 161
    Summary 164
    Discussion Questions 165
    Assignments 165

**9 Selecting an Innovation Project: Projects that
Add Lasting Value 167**
    Introduction... 167
    Step 1: Assessment 169
    Step 2: Normalize and Nominate 170
    Step 3: Objectify and Operationalize 174
    Step 4: Validate and Verify 178

Step 5: Adaptation and Alignment 185
    Acceptance of Change 188
Step 6: Tabulate and Track Performance 190
Step 7: Execute and Evaluate..192
Summary 193
Discussion Questions...194
Assignments 194

10 Leading Innovation Project Success: From Concept to Reality...197
Introduction...197
The Impact of Innovation..198
Macroinnovation 200
Microinnovation 202
Implementing a Successful Project..................................202
    Scenario 1: Incremental Improvement.......................203
    Scenario 2: Replacement.. 214
Summary 223
Discussion Questions...225
Assignments 225

11 The Federal Government: 2015 and Beyond........227
Introduction...227
Summary 234

References 235

Index 239

Acknowledgments

As I always do, I want to thank my dear wife, Heidi, for her understanding and selfless support of my professional and spiritual activities. You are my world and you have my love forever!

Frank Voehl has been an extremely valuable resource for me in formulating this book. He helped greatly with the assessment, process, and strategy pieces. I was able to modify, with permission, the IQ (Innovation Quotient) instrument (created by James Higgins) from its present state to that for assessing N²OVATE™ preparedness. I want to thank Frank for his immeasurable support and for his continuing contributions.

Jim deVries has also been an excellent resource and a great professional colleague.

Finally, Dr. Buzz Kennedy has been a great friend and a person I trust and believe in. This is only the beginning of a great professional relationship!

Dr. Gregory McLaughlin

I wish to acknowledge those individuals who were instrumental in helping complete this book. First and foremost, to my partner and champion, Jannette (my partner of 36 years), whose love and invaluable knowledge, timely insight, and unwavering encouragement made this book a pleasure to write. To my children, Darell, Samantha, William (Will) and Shea. I'm not only proud of you all, but you mean the absolute world to me, and knowing your love and support would always be there regardless of my contributions to humankind leaves me in the best position a Dad could ever hope for.

To my brothers at the United States Air Force's Air Mobility Command (AMC): Lt Col (Ret) Anastacio "AL" Lambaria, my closest friend and long-time mentor. Your years of devoted friendship, sage advice, and continuous support when I left a 30-year career serving this great Nation to pursue my passion of sharing knowledge with others will never be forgotten. To my friend and innovation visionary, Lt Gen (Ret) Robert "Dice" Allardice, who shared his wealth of knowledge, experience, and insight honed from years of implementing successful game-changing initiatives that modernized the United States Air Force military culture. To my good friends SMSgt (Ret) Bradford "Brad" Tague, Lt Col (Ret) William "Bill" Dalonzo, and Lt Col (Ret) Matthew "Matto" Roller, who all took me under their wings and willingly shared the world-class knowledge one can gain only through years of direct involvement and selfless dedicated service to the American public, I owe you all a great debt of gratitude, my brothers.

To Captain Jeremy Baker, an exemplary young leader, Defense Acquisition guru, and devoted family man, your vision and drive to always ask the hard questions regardless of the audience was simply a breath of fresh air in the government bureaucracy.

Finally, to my dear friend and mentor, Dr. Greg McLaughlin, the opportunity to develop new innovation processes and solutions approaches adaptable to rapidly changing environments presents some unique opportunities not only for the private sector, but for the government, which can benefit from agile and flexible fact-based processes to deliver actionable and reliable information to the right decision maker at the right time.

Dr. William Kennedy

About the Authors

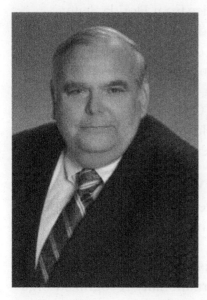

Dr. Greg McLaughlin is a managing partner at Innovation Processes and Solutions, LLC. He brings a broad set of technical and practical expertise in quality improvement, innovation, and data analysis. Beginning as an analyst, he progressed quickly to director of research at a Fortune 200 company. Refining his skills in continuous quality improvement, he worked for Dr. W. Edwards Deming as instructor/consultant. Greg authored a book for research and development in organizations (*Total Quality in Research and Development*, 1995) committed to quality improvement. He was an early adopter of Six Sigma and worked many years as a Six Sigma Senior Master Black Belt, saving organizations over $300 million. Many projects resulted in innovative products and services. His most accomplished skill is in interpreting data and finding a practical application. He can look beyond the numbers to find a solution to complex problems.

Dr. McLauglin's skill set organically transitioned into developing innovation strategies, deployment, and sustained success, as evidenced by the creation of the ENOVALE™ and N²OVATE™ methodologies. He maintains a leadership role in developing training, tools, books, and publications for both practitioners and scholars. His latest publications are *Chance or Choice: Unlocking Innovation Success* (2013); *ENOVALE™: How to Unlock Sustained Innovation Project Success* (October 2013); *Leading Latino Talent to Champion Innovation* (2014); and *Unlocking Sustained Innovation Success in Healthcare* (2014).

His educational achievements include a doctorate in business administration from Nova Southeastern University, a master of science degree in statistics from Florida State University, and an undergraduate degree in meteorology from Florida State University. Greg was director of doctoral research at Nova Southeastern University and was instrumental in creating an innovative dissertation process for the doctor of business administration degree at Capella University. Since its creation, the DBA program is the largest and most profitable doctoral program in Capella University history.

Dr. William R. Kennedy is currently a primary managing partner at Innovation Processes and Solutions (IPS) Consulting. He is an award-winning and internationally recognized organizational leader and author with over 30 years experience in the public and private sectors. Buzz has an extensive background in leadership and management in information technology, defense, and aerospace acquisition, requirements, and program management. He is also considered a subject matter expert in strategy and innovation development, manufacturing and production, reliability-centered maintenance, Lean manufacturing, logistics and supply chain management, and training program development, design, and instruction.

He is also a highly decorated US Air Force veteran with a global perspective and more than 20 years of international experience across the Far East, Middle East, and Europe. His diverse background and natural ability in developing tailored innovative solutions and high-performance teams have led multiple organizations in achieving world-class performance.

His educational achievements include a doctorate in business administration from Capella University, a master's degree in secondary education from Grand Canyon University, and a bachelor's degree in business management from the University of Maryland.

Chapter 1

Why This Book Is Necessary

Introduction

In the late 1990s, the US government's leadership was driven by the great epiphany—the dawn of the information age—that set in motion the creation of a new virtual frontier driven primarily by the advent of the World Wide Web. This new frontier took on the name *cyberspace*, and its impact on how we live our daily lives is immeasurable. This new cyberworld had a dramatic impact on national security and the world of business, redefining the very fabric of the global business landscape and trends toward globalization and cross-border integration. Stretching far beyond the boundaries of business, the information age became the catalyst for the knowledge age, driven principally by innovation and the insatiable appetite for unfettered access to unlimited information at any point or any time in the world. Subsequently, the need to effectively harness, manage, monitor, and secure the rapid evolution of this infocentric frontier and the virtual global information grid has involved a monolithic effort that remains a chief concern for the nation's security and economic well-being. Beyond the immediate concern of national security and the government's increasing reliance on and proliferation of information technology (IT) is the need to effectively manage the daunting amount of information and knowledge available on the grid.

As the federal government came to grips with the implications of the cyberworld, it became evident the global rules of engagement regarding access to sensitive information and knowledge generation had changed. The United States was compelled to take a leading role or risk losing strategic, operational, and perhaps even the tactical advantage they had worked so hard to achieve since the end of the Cold War. To assist in sorting through these matters, the government created the Federal Knowledge Management Working Group (FKMWG) and established the position of chief knowledge officer (CKO) in several federal agencies. These two key steps were instrumental in establishing the initial groundwork for synchronizing several important government agencies (FKMWG, 2011). Sequestering inputs from a number of subject matter experts across the government, the FKMWG built several living documents to help set the tone and frame the government's initial strategy. Chief among those foundational documents was the FKMWG charter, which spawned federal policy and guidance for integrating innovative knowledge management (KM) principles into their overarching strategy and daily operations.

With the government's increasing commitment to combating terrorism, maintaining world peace, and engaging in the hegemonic pursuit of a global democracy, the increasing reliance on KM methodologies has necessitated continuous process improvement (CPI) and advanced systemic requirements for ensuring national security, global information dominance, and protection of free-market trade and economic stability. A generally accepted catalyst for innovation, the federal community has made significant investments in IT, spawning culture-changing improvements and a host of innovation opportunities.

In today's knowledge-driven global environment, fueled by an ever-increasing appetite for timely information through seamless, robust, and reliable technology and continued innovation, decision makers and senior leaders across all government agencies are seeking new ways to increase efficiencies. In our work, we introduce our transformational innovation approach (N²OVATE™), the tools, and processes that we believe can make significant improvements in how the federal sector handles innovation opportunities.

Facing austere times, fiscal prudence, sequestration, and deep personnel cuts have become driving forces in changing traditional strategies and philosophies toward daily operations. Further, in an effort to scale back an overinflated government plagued by years of bureaucratic inefficiencies, political fragmentation, and a behemoth Napoleonic organizational structure, the government must now make some hard decisions and prioritize the nation's requirements. In supporting this effort, the public sector is exploring private-sector business practices for feasibility and adoption.

While the government has squandered trillions of dollars on a plethora of questionable defense, social, and economic reform programs over the past four decades with mixed results, some might argue the private sector has enjoyed an extended period of prosperity and success. Generally defined by successful returns on investments (ROI), innovative growth continues to drive significant improvements (e.g., intellectual, structural, and human capital) and competitive value for shareholders, stakeholders, and the global community in general. In reality, the private sector is flourishing while the public sector is in decline. This stark contrast between the public and private sectors has placed the government in a precarious position where change is no longer a consideration but a mandate. Creating effective strategies and methodologies will empower the organization to make innovation a priority and mandate. This is the primary reason for writing this book: to provide a blueprint for success.

In summary, *effective innovation* is a term used frequently to describe the process of selecting new ideas to develop into new products or new technology. It is a definitive way to soar above your competitors when competition is an issue. It is a "buzzword" for companies and organizations to suggest technical competency. Every organization wants to be innovative but attaining this level of achievement seems to be difficult. This book helps explain and introduce the process or strategy to achieve that goal, especially when dealing in the federal sector.

Why Innovation Is Such a Difficult Goal to Achieve

Innovation, as an agency or organizational goal, seems to be a reasonable objective. Yet, the execution and accomplishment of such a goal has proven elusive for many organizations. Our research demonstrates that individuals perceive (understand) innovation in numerous ways. Whether something is innovative or not is a matter of meeting a strict definition or requirement. Innovation is truly in the "eye of the beholder"; it exists because it has met some criteria or achieved some objective that the individual established prior to using or purchasing the innovation. As diverse as the population is, so are the criteria people use to judge an item (product, technology, or service) as innovative. This judgment individual's exercise is a set of interrelated experiences, acquired knowledge, and satisfaction with similar products in the past. The diversity of opinion, knowledge, and experience provides an entire chapter in this book. The next chapter addresses how to deal with this issue of understanding how individuals perceive (recognize) and define innovation. Understanding how innovation is defined and recognized brings it closer to a reality for the organization. That reality can be of major benefit to the organization in the value that it brings both internally and externally.

Ongoing Research

In Chapter 2, we provide a brief diagnostic review of a recent survey of how members of the online, social networking community, the FKMWG, view innovation from the individual level. Historically, research efforts in both the private and public sectors have focused on individual perceptions of innovation and have centered on cultural and generational factors as identifiable variables. By design, the FKMWG study (2011) sought to broaden the definition of culture to one beyond national origin to that of a professional community or organization associated with federal activities. The study also added the elements of generation (age), gender, and job function or classification and was the first of its kind.

We feel there is a relative linkage between how federal KM professionals perceive innovation (at the personal level) and how the remainder of the federal workforce perceives innovation in general. We do not believe the results of this survey are definitive of every federal community and culture, but we feel it is a starting point for agencies and organizations to begin the journey toward an effective and efficient innovation culture.

Far too often, determination of innovation project roles in the federal community depends on position versus knowledge and skill sets germane to the overall goal of a specific objective or desired outcome. Coupled with the information we share in this book, the potential exists for the study's results to provide a foundational starting point in identifying common elements essential in aligning talent, selecting team composition, and deciding agile leadership roles and approaches for product, process, and service success.

Culture and the Environment

Culture also plays a role in defining how to recognize and benefit from innovation. When speaking of culture, the word is often confused with ethnicity. *Ethnicity* refers to a person's natural biological affiliations (language, skin color, heritage, etc.). *Culture* refers to the influences affecting the individual in regard to making a decision, namely, that of purchasing or using (recommending) the innovation. These influences can come from national origins (where a person has lived), the organizations that employed the individual, and membership organizations (unions, professional societies, philanthropic societies, etc.) and affiliate influences (family, peer, and governmental organizations). All these influence the individual and the individual's decision making.

The environment also contributes to the recognition of innovation and subsequent decision making regarding an innovative item. The environmental component is more difficult to classify as it is situation dependent. The economy and its impact have a great influence on innovation. This is especially true with the business community, which also influences government. Internally, an open and collaborative environment supports innovative practices and policies. This includes balanced and effective communications, feedback, and directives. Overall, a conducive and receptive environment supports a rapid acceptance of innovation into the organizational culture.

Dynamic Capabilities

Building on the environmental issues, the organization must utilize its own dynamic capabilities to achieve sustained and successful innovation. In a broad context, dynamic capabilities are the routines (procedures or processes) performed at varying levels within the agency or organization to achieve the assigned responsibilities. At the organizational level, it is a high-level routine (or collection of routines) that, together with implementing input flows (inputs and outputs from other processes), provides the agency's leadership and management (decision makers) with an objective view or set of decision options (actionable information) for producing significant outputs to meet a particular goal or set of objectives. They key here is objective, fact-based actionable information free from personal feelings and motives.

Further, dynamic capabilities are those skills needed within the agency or organization to build internal and external competencies in an integrated workplace conducive to a changing environment. In other words, these are skills that determine an organization's ability and propensity to "integrate, build, and reconfigure internal and external competencies to address rapidly changing environments" (Teece, Pisano, & Shuen, 1997, p. 516). Federal agencies capture dynamic capabilities in the form of a standing or standard operating procedure (SOP). SOPs are employed to capture the unique processes and procedures in a variety of contexts throughout the healthcare, aviation and aerospace, military, education, and manufacturing and production communities. Chapter 4 dives deeper into how dynamic capabilities are developed and improved.

Requirements

To begin the innovation process, the organization (or individual) must define why the innovation is required or wanted. There must be a reason for the organization to embark on such an effort. In an environment that requires little change or one not threatened by advances in technology or user demand, innovation is often dormant. Events, situations, or requirements can generate innovative activities. In all cases, a need or specific requirement generates the innovation.

Innovation begins with a need (requirement) that is presently unsatisfied or unfulfilled. Requirements can generate innovative events, especially if these are contractual or product specific. The key word is *unsatisfied*: It is something someone or something (an organization) does not have but does need. Most needs have an external focus, but often an internal need takes precedence. What drives innovation (makes it a reality) is this set of needs or requirements.

We discuss how to collect, analyze, and evaluate these needs (requirements). If the need or requirement cannot meet these criteria, then it will not become an innovation project. Further, unlike the prescription N²OVATE™ offers, some might argue that some programs within the requirement (need) structure do not have the right mix of players or keep the right players involved in an innovation through the entire life cycle based on their own succession plans. Based on personal knowledge and experience, there are some acquisition programs within the federal community that require a stable, experienced, and trustworthy core of professionals; one of those is Presidential Aircraft Recapitalization (PAR) program.

As a point of reference, whatever Air Force–owned and –operated fixed-wing aircraft in which the president flies is referred to as Air Force One. The aircraft the vice president flies in is referred to as Air Force Two. The PAR program focuses on replacing the fleet of two aging Boeing 747-200 wide-body aircraft, known as the VC-25 (tail numbers 28000 and 29000) when the president is not aboard; these aircraft flew their first missions in 1990. Factoring in the actual date these two 747-200s became a presidential support platform, they are in their 25th year of service at the time of this writing. It stands to reason, considering economics and past practice, that logistical support challenges in tandem with cost and performance are motivational factors for an innovative solution and recapitalization.

As the VC-25 community is a tight-knit group, several common threads are defining factors that separate this program from many others within the federal ecosystem. Several of the key players and program managers in both the public and private sector have been associated with the platform for many years. In other words, the owners of this capability trend toward continuity of purpose by retaining top-shelf talent in the program. Innovations are introduced and acted on through fruition, and the influence of transitional leadership and program managers is mitigated through this practice and policy. Not to relegate military and civilian leaders who have played major roles in the program to lesser positions, a matter of importance and national security dictates an individual's influence. Now, if innovation at all levels within the federal government were handled in this manner, one might wonder what the outcome would be for other innovative ideas.

Assessment

Hand in hand with establishing needs and requirements is that of assessment. *Assessment* refers to the collection, analysis, and interpretation of surveys, focus groups, and related data. Assessment permits an evaluation of needs and requirements to determine the best course of action for an agency, organization, or business. Evaluations also include an organizational assessment, Innovation Readiness Index, innovation strategy, how employees view innovation, and environmental and value assessments, to name a few. The assessment permits the development of a unique and focused innovation process and solution set tailored to the needs of the organization.

Assessment leads directly to diagnostics. Diagnostics provide the mechanism to establish a focal point at which innovation can succeed. By surveying the organization, we create an overall evaluation of the organization, its highlights, and its potential obstacles to innovation. We summarize the data and then present the results to management and discuss its implications. This information is a starting point, a place to begin our conversations.

Strategies for Innovation

Assessment results lead to a better understanding of the innovation strategy in place within an organization. That strategy (if there is one) is crucial for sustained success and must be visible throughout the organization. Without a strategy, innovation occurs more by chance than by choice. In the best situation, an innovation strategy consists of three components that usually occur at the tactical, operational, and strategic levels within an organization. In fact, a "one-size-fits-all" approach does not work well with innovation because at the strategic level the focus is on decision making, and implementation is the focus at the tactical level.

This book reviews existing innovation strategies for overall impact and effectiveness. For those organizations without an innovation strategy, we discuss a method to create a strategy. Each organization will require a unique set of strategies to succeed. Because the most desired outcome is to identify and execute innovation projects, this book highlights these strategies. We present a core tactical strategy that is modifiable for each agency or organization.

In addition, we subdivide projects into large-scale (macro-) and smaller-scale projects (microinnovations). This permits innovation at the individual rather than just the functional level. Regardless of where an individual fits within the organization, innovation projects can bring value to the individual or to their departments or groups. More innovation projects create more value and benefit to the organization.

Finally, we provide a discussion of success factors associated with innovation. Success factors (sometimes called key performance indicators [KPIs], etc.) are metrics developed to identify, track, and audit the success of innovation projects. The development of these metrics coincides with the development of the project's objective. These measures define success and value to the customer (user), employees, and management.

Innovation Process Management

Organizations need innovation projects to return value to the bottom line. Therefore, our main strategic thrust was to develop a process that could sustain a successful outcome. First, we developed a seven-step process, called ENOVALE™, to select the best innovation projects for implementation (McLaughlin & Caraballo, 2013). This series of seven steps can lead a project to the value creation phase. Next came the development of a different (more focused) version of ENOVALE™, which was used to implement the project. This book updates the seven-step project selection process by generalizing it to any business or organization (Chapter 9). The new process, renamed to N²OVATE™, is more adaptable to varying types of organizations and businesses. Chapter 10 discusses one type of implementation strategy, as applied to the case study from Chapter 9. We expect that readers will find this a new (and innovative) process that is universal in nature and complimentary to their goals for achieving value in the existing innovation strategies.

The first installment of innovation management is in selecting the best project based on the N²OVATE™ criteria. We use a federal-sector case study example to describe this process of evaluating a project for its overall value and applicability. The process begins by highlighting a need or requirement that will drive the innovation. If the need (requirement) is capable, sustainable, and viable, the project moves to the next stage. This next stage establishes the project objectives, functional requirements, assessment of risk, and limitations and assumptions. The remaining stages focus on alignment to the objective and linking the project to financial performance. Acceptance of a project means that it can proceed to implementation (be executed by the organization).

Chapter 10 discusses a common strategy to implement an innovation project in the federal sector. The process

(N²OVATE™) adjusts to the type of innovation required. The case study from Chapter 9 becomes an example for one common type of innovation outcome. As an alternative, a contrasting process (with a strategic focus) is compared to the more common approach. This demonstrates the power of N²OVATE™ to operate at multiple strategy levels and be an effective innovation strategy. Rather than hunting for a strategy for the federal sector, a modified version of our innovation process management is available. Development of the strategy (process) permits an organization to achieve any desired outcome. We know of no innovation strategy that is as flexible and dynamic as N²OVATE™.

Incremental Innovation

Finally, incremental innovation (Northwestern Ontario Innovation Centre, 2015) begins with the process of defining requirements and needs and ends with value-added outcomes. It is truly a process (Figure 1.1) of evolving the organization, its stakeholders (customers, users, and suppliers), and its employees. Chapters 2–10 describe the process in detail. The flowchart in Figure 1.1 defines the shift in management thought and action that must accompany this change. This need to reassess the organization and its embrace of an innovative culture becomes the backdrop of the book. Becoming an innovative organization takes time, effort, and determination. Transforming the organization is an evolutionary process requiring the support and resolve of management to achieve this goal.

Although all that must occur seems like a daunting set of activities, management can incrementally implement these items over time. In fact, most clients begin with project selection and implementation before attempting changes in culture, decision making, consensus building, and collaboration.

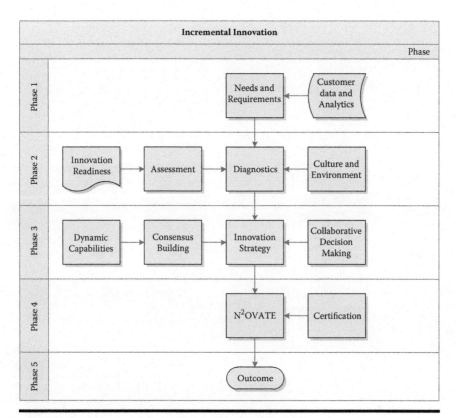

Figure 1.1 Incremental innovation process.

Summary

This chapter introduced the book to the reader. Rather than just providing a dialogue, this book contains numerous process flowcharts, tools, and strategies for innovation within an agency or organization. We have presented why the book is necessary, the complementary nature of diversity and innovation, influences on innovation, and the power of innovation. We have also walked the reader through the innovation processes and solutions created to understand what initiates incremental innovation and to determine if the organization is ready for innovation. Finally, we discussed how to take a project to a successful conclusion.

Figure 1.1 Information initialization process.

Summary

Chapter 2

Innovation: A Combination of Perspectives

Introduction

Innovation is both a concept and a reality. It often begins with an idea, but can translate into a product or service. We know innovation when we experience innovation. Individuals use their perspectives to evaluate and judge innovation. The paradox is that we may "experience" it differently from someone else but arrive at the same conclusion. These perspectives are as diverse as those who experience the innovation. Yet, these perspectives come together or coalesce to explain the concept of innovation. This "coalescence" of perspectives provides a wellspring of opportunity.

Perspectives develop from a culmination of an individual's knowledge and experience. Perspectives form a frame of reference on how an individual assesses or evaluates innovation. Individuals use a variety of sources of knowledge and information before deciding whether a product, service, or process is innovative. The difficulty is in understanding which

15

perspectives each individual employs and the degree or depth applied to label an item "innovative." The discussion focuses on the five perspectives known to influence the evaluation of innovation: opinion, satisfaction, choice, knowledge, and judgment. These perspectives come together (coalesce) to evaluate whether a product, process, or service is innovative.

This chapter examines and describes the many diverse perspectives that individuals use to judge innovation. These issues range from defining innovation to individual satisfaction. The individual's urgency of needs drives innovation in almost every field. Different people need different forms of innovation—this difference is the reality in describing the concept. Understanding this coalition of perspectives is the crux of understanding innovation.

Coalescence/Combination

Five forms of coalescence, as it applies to innovation, are described in this chapter. Figure 2.1 details five unique components (forms) (opinion, satisfaction, choice, knowledge, and

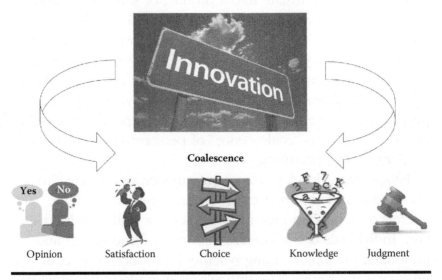

Figure 2.1 Innovation coalescence.

judgment) that coalesce to describe an individual's understanding of innovation. This chapter describes each component and follows with a discussion of how these coalesce to describe innovation.

Opinions: Expanding the Definition of Innovation

To begin, gather five people in a room and ask them to define innovation; expect a number of unique and often-disparate responses. Some will say that only new technology is innovative; others will say it is incremental improvement; yet others may voice that only discoveries and inventions are innovative. Each person individualizes the definition to what they know and what they experience. For some, innovation must change their lives to be relevant. In fact, some may not recognize it or simply dispute its advantages.

Level setting the definition of innovation is critical because without a definition or description, innovation could become like "beauty": purely in the eye of the beholder. Consider that researchers working in innovation have identified over 60 different definitions of innovation (Bareghreh, Rowley, & Sambrook, 2009). Therefore, we contend that defining innovation as a single definition is far too elusive for this book (but we are working on it). Thus, in this book we discuss innovation from the perspective of how it is understood or perceived from a more tangible perspective and approach.

Many consider that innovation begins with an idea, such as that associated with an invention. Ideas alone may not meet what is required or wanted from the consumer, the patient, or the public. Ideas that meet needs (wants or desires) may be a better choice for innovation because there is a dedicated desire for some product, service, or technology. If there is a need, someone will find a way to meet that need. Innovation, then, occurs when a product, service, or technology meets or exceeds an unfulfilled need. That said, we argue that innovation is more complex than just fulfilling a need or requirement; it also has a distinct tangible nature. The tangible piece relates to satisfaction, performance, and use of the item. Understanding this nature of innovation leads us to a better definition of the phenomenon.

If an individual encounters (uses) a product for the first time or has limited or no experience with the item (product, technology, or service), it is considered "new." New items are often portrayed as innovative because the items never existed previously. Yet, this restricts the definition of new to only inventions and discoveries; we also feel that the word *new* can describe a new application or new approach. The iPad is an example of a new application. The iPad may contain innovative parts or software, yet it is an improvement over an existing PC (personal computer) tablet. This does not diminish its innovativeness—it just classifies the new status. One final form of new is a new approach. Here, an existing product is updated, reformulated, or repositioned. It satisfies a need, but competitors can easily replicate it or provide a suitable substitute. The life cycle of a new innovation is dependent on its originality: If the design is unique or original, the life cycle will be longer.

Based on ongoing and replicated research (McLaughlin & Caraballo, 2013b), three observed outcomes of tangible innovations are new, improved, or changed. That is, the individual can classify the innovation as unique or new, improved, or substantially changed by comparing it to previous encounters (experiences). This definition does not limit innovation to new products, services, or technologies. By expanding the traditional definition, innovation is possible in diverse applications and environments. One such area, traditionally overlooked, is services, rarely categorized as innovative. Because technology does not drive most service innovations, many feel it does not qualify as innovative. Yet, a new service may bring numerous value-added opportunities. Opportunities become more frequent and innovation more of an expected or desired outcome. This coming together of opportunity permits a more inclusive definition. Innovation is more than just the newest fad or latest gadget: It has become a strategy for the organization to improve.

One tangible aspect of innovation is performance. Performance of an item must exceed expectations to be innovative (as well as satisfy a need). When an organization decides on an innovation improvement strategy, improving performance (value) becomes a tangible measure of success. Performance is a direct measure of outcome or accomplishment. Most organizational functions monitor their performance in some manner. Traditionally, the operational side of any organization easily monitors performance, measuring such qualities as efficiency and effectiveness. The military monitors readiness or preparedness, which is a measure of performance. Yet, service and human functions often measure performance in less-quantifiable methods. For example, an employee or associate's yearly performance evaluation is often subjective, lacking many measurable standards. Asking an employee to improve task performance when the information provided is more opinion than fact is ludicrous. Poor or nonexistent measurement specifics prevent lasting performance improvement.

Measuring performance requires establishing a standard from which to measure future performance. This permits incremental improvement—because management (leadership) can track performance (as it relates to improvement).

Incremental improvement (which can occur at the organizational, departmental, or individual level) is best "fit" for the organization, as it is a concerted (orchestrated) effort. Like continuous improvement (which has a process orientation), incremental improvement has a broader focus, including policies, procedures, process, and programs. It is the essential innovation strategy for the twenty-first century. For the government sector (federal, state, and local), it is a natural innovation strategy. Incremental innovation (improvement) is strategic in content versus the one-time, game-changing innovations affiliated with a blue ocean or disruptive innovation opportunity. These types of innovation literally change the existing paradigm—and are correctly classified as game changers. We do not discourage these approaches but realize their limitation when considering a long-term approach. Incremental innovation is an integrated approach; it can be organized, managed, and measured. This differs also from continuous improvement because not only must performance improve, but the innovation must satisfy a need. For the government sector, incremental improvement will provide the best long-term strategy.

Finally, innovative change is change (replacement) with a positive outcome. Change occurs through planning and leadership support. Similar to change management, the focus is on the decision made and its consequences. An example of innovative change is renewing a driver's license. In many states, drivers can renew their license online without the necessity of waiting in long lines (queues). Policies and procedures changed, replacing the need to come in to the state office (a previous policy). It may be an improvement as well with the elimination of wait time (a performance measure). The number that recognize this as innovative is unknown (given the large number of opinions), but it certainly meets the criteria.

We suggest that the organization level sets the definition of innovation to the three categories, depending on the situation. Reaching consensus on the definition of innovation may seem trivial, but it aligns individuals to the same goals and objectives. Once aligned, the implementation team will achieve consensus and collaboration faster and with less dissention or diversion. This has been our finding with previous groups.

A Research Study

A research study reported by Kennedy (2014) focused on eliciting personal perceptions of innovation from an online social networking community associated with the federal government, such as the Federal Knowledge Management Working Group (FKMWG). Historically, research efforts in both the private and public sectors have focused on individual perceptions of innovation and have centered on culture and generational factors as identifiable variables. By design, the FKMWG study broadened the definition of culture to one beyond national origin to that of a professional community or organization. The study also added the elements of generation (age), gender, and job function or classification. This study built on the work of McLaughlin (2012), Caraballo and McLaughlin (2012), and Zhuang, Williamson, and Carter (1999).

The central research question for this study concentrated on how members of the FKMWG perceive innovation at the individual (personal) level. Selecting an identifiable group such as FKMWG membership presented distinct opportunities to expand the innovation management body of knowledge (IMBOK) by exploring a new homogeneous group, the knowledge management (KM) professional. We based the central discussions in this chapter on a diagnostic approach: how the data was collected, our assessment of the resulting data, and how we interpreted the data presented in this study (Kennedy, 2014).

Background

The FKMWG study covered a 26-month period (July 2011 to August 2013), focusing on discovery of how the FKMWG members view innovation from an individual or personal perspective. The central research question in this quantitative, nonexperimental study was, "How do members of the FKMWG perceive innovation at the personal or individual level?" The central question was augmented by the following demographic and function-related research questions and explorations:

- How does gender affect a KM professional's view of innovation?
- How does generational cohort affect a KM professional's view of innovation?
- How do functional association or job classification affect a KM professional's view of innovation?
- The addition of the second-order research questions (e.g., generation, gender, job function or classification) broadened the IMBOK and opened new opportunities for further research in how gender, generational grouping, professional association, and job classification influenced views on innovation on a personal level.
- Further, the study focused on building a better understanding of how innovation diffuses over a period of time (Figure 2.2). The study's survey was voluntary and anonymous and targeted the 650 members of the FKMWG.

The four variables studied (age, birth data, professional association, and job classification) are detailed in Table 2.1.

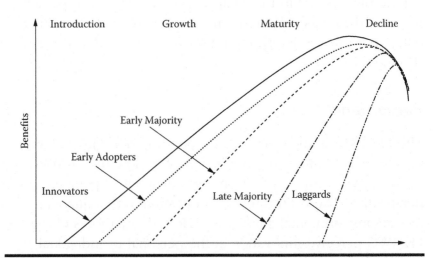

Figure 2.2 Knowledge management system rate of diffusion. (From Kennedy, W. R. Individual (Personal) Perspectives on Innovation: Federal Knowledge Management Working Group. PhD diss., Capella University, 2014. ProQuest, UMI Dissertation Publishing [3611870].)

Table 2.1 Research Study Variables

Question and Variable	Category
a. Gender	Male
	Female
b. Date of birth (age)	Veteran: born prior to 1925–1942
	Baby boomer: born 1943–1960
	Generation X: born between 1961 and 1979
	Generation Y: born between 1977 and 1992
	Generation M: born between 1992 and 2010
	Generation Alpha: born after 2010
c. Professional association	Military (public sector)
	Government civilian (public sector)
	Contractor (private sector)
d. Job classification: personal perception (nature) of daily contributions	Technical in nature
	Nontechnical in nature
	Contributions vary based on project/assignment

The results of the study, displayed in Figure 2.3, indicate the strength factor of each demographic group's perceptions. For all groups, the strongest innovation was either improve or improve and change. New and change alone were rated but less important. This result complements the ongoing research and highlights the need for these strategies in the federal sector.

Summarizing the Results

Understanding how KM professionals perceive innovation at the personal level provides federal agencies a foundational

Position	Demographics			Strength Factor		
	Generation	Gender	1	2	3	
		Female	Improve and Change	New		
		Male	Improve and Change	New		
	Generation X		Improve	New	Change	
	Baby Boomers		Improve and Change	New		
Profession 1 (Military)			Improve and Change	New		
Profession 2 (Government Civilian)			Improve	New	Change	
Profession 3 (Private Sector Contractor)			Improve and Change	New		

Figure 2.3 Research study results, FKMWP.

starting point for identifying common elements essential in aligning talent, selecting team composition, and deciding agile leadership roles and approaches to products, processes, and service factors for improvement projects. Too often, position versus knowledge determines innovation project roles in the federal sector with skill sets germane to the overall goal of a specific objective or desired outcome. Much like managing major acquisition programs within military organizations, the assignment of innovation efforts is to the ranking individual who has "time in the cockpit" and can parlay the "voice of the operator." In some cases, these professional leaders often lack the appropriate skill sets and knowledge base of how to run an innovation event. Further, they are often the one who assigns team members the key support or decision-making roles without a clear and pronounced understanding of expectation or the innovation process outcome.

Innovation requires more than perceptual acknowledgment; it must please the customer or user. Individuals want the item to satisfy not only their needs but also their wants and desires.

Understanding the concept of satisfaction, as it contributes to innovation, is critical.

The Contribution of Satisfaction

Needs (Requirements) Satisfaction

For innovation recognition, not only must the consumer or user be happy with the product, service, or technology, but they must also satisfy an outstanding need, want, or desire. Otherwise, the product, service, or technology is merely an improvement from previous versions. Often, the word *need* is synonymous with the word *requirement*. A requirement is an outstanding (unsatisfied) need that provides a distinctive competitive advantage (value opportunity). Individuals certainly use criteria for deciding whether a need is truly satisfied. These criteria may be as simple as the following:

1. The need provides value.
2. I want the item (value), so the need is satisfied.
3. An outstanding situation requires the item (value), satisfying the need.
4. My needs have been satisfied in the past (value) by like (or similar) innovative items.
5. There are no substitutes (need is unsatisfied).

Figure 2.4 describes a flowchart approach to the process of satisfying a need or requirement. Triangles are decision points, boxes action items. After identifying a need, the individual must ask what value there is in pursuing this need. If the need is critical, then the individual will seek to satisfy the need or find a substitute. Of course, the option exists to do nothing or to wait for or to propose an alternative.

Requirements that define parameters, conditions, or contractual agreements (as in legal agreements) drive behaviors

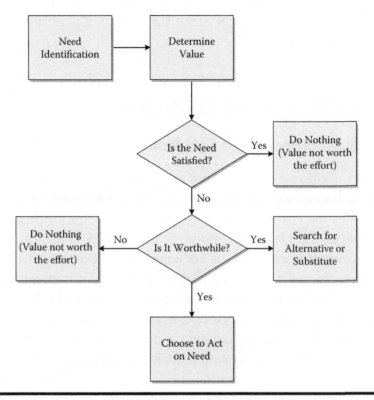

Figure 2.4 Identifying and acting on needs (requirements).

in similar ways to needs. These requirements are often conditions that must be satisfied before receiving approval (which is a value). Innovative approaches to meeting these requirements can generate tremendous value.

After identifying a need (requirement), the next step is to evaluate the need. Evaluating these needs (requirements) determines their viability and predictive power. We recommend three unique characteristics to assess needs: sustainability, capability, and viability. These are defined as follows:

1. Sustainability: Criteria that assesses the length of time the need is competitive (adds value).
2. Capability: Criteria that assess performance and customer satisfaction issues.

3. Viability: Criteria that assesses feasibility and practicality (McLaughlin & Caraballo, 2013a,b).

Once the analysis is complete, the information greatly enhances the ability to predict an innovative outcome. Satisfying these criteria validates the need (requirement).

However, satisfaction of the need alone is insufficient. Innovation begins with answering a need, but that need may quickly become irrelevant in a highly interactive environment. A viable need must also satisfy the customer or user. Therefore, it is important to capture the user's perceptions of satisfaction, given this dynamic and changing environment.

User (Customer) Satisfaction

Satisfaction is the difference between what a person expects and what a person actually encounters (or experiences). A customer or client may be satisfied but decide not to change their purchase behaviors or acquisition behaviors, even if recognizing the innovative attributes of a product or service. Satisfaction factors may alter a decision that seems rather straightforward. For example, people select airlines based on price rather than previous experiences. This is the reason that airlines continue to cut services, such as leg room, while increasing fees. Their models tell them that pricing is the key determinant. These factors will also affect innovation recognition and purchase.

By combining perceptions and expectations, individuals judge whether a product, service, or technology is innovative. Customers use their experiences and knowledge to evaluate an item. Customers, clients, or the public use widely diverse tools, personal experiences, and information gathered on the product, service, or technology. One key measure that greatly influences a decision is that of performance given that

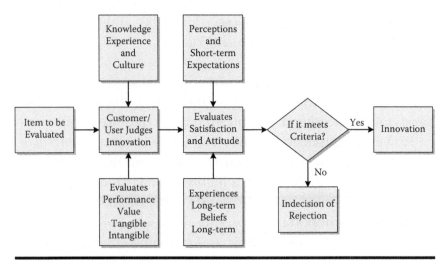

Figure 2.5 Innovation evaluation and satisfaction with the decision.

it provides multiple judgment criteria. When performance (greatly) exceeds expectations, then individuals perceive the outcome as innovative. The word *greatly* varies from individual to individual.

In addition, longer-term measures of acceptance and use, such as attitudes, are critical as well. Individuals with experience and knowledge of an item will use their experiences and beliefs to evaluate an innovative item. Therefore, individuals will rely on their knowledge and experiences with like or similar products, services, or technology to judge value and possible acquisition. Figure 2.5 illustrates how an individual evaluates innovation from initial encounter (innovative outcome) to decision. It is a complex process involving knowledge, experience, performance, and judgment. (Many of these attributes come together to define [explain] innovation.)

Inevitably, the consumer must recognize the item as innovative and decide whether to purchase (acquire) the item. Purchase or acquisition decisions involve many intangible criteria, unique to each individual and situation. One key criterion is satisfaction and its antecedents. Satisfaction involves both expectations and perceptions, yet the individual often considers related factors, including

1. Loyalty
2. Emotional attachment
3. Preferences
4. Personal values
5. Pricing

These elements (factors) of satisfaction contribute unequally (Figure 2.6), adding to the complexity. Often, indecision is due to competing factors. To predict an accurate outcome, the organization must understand how the decision process functions, account for varying factors, and realize the individual nature of the decision. Hence, this is the reason it is so difficult to predict whether an individual identifies innovation as such.

A person could recognize innovation but decide that the need is not strong enough to require purchase/acquisition. Minor factors affecting that decision may include tangible items such as physical characteristics and functionality and the intangible characteristic of desirability. For governmental organizations or agencies, no product, process, service, or technology is guaranteed to be recognized as innovative by all users. What is critical is the value-adding properties that innovation generates.

Satisfaction is a short-term perception. Numerous good or bad experiences will reinforce attitudes and beliefs.

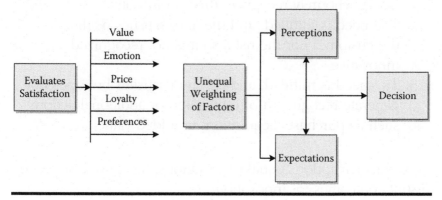

Figure 2.6 Elements (factors) of satisfaction.

Longer-term attitudes and beliefs influence purchasing decisions (value assessments). Satisfaction contributes to innovation, especially when considering the choices consumers and users make routinely. Disregarding or deemphasizing this concept further degrades our understanding of innovation.

The Amalgamation of Choices

Building on the previous section, identifying an item as innovative involves both perceptions and expectations. When perceptions do not meet expectations, then there are three distinctive outcomes for innovation:

1. The need remains unsatisfied, and there is neither satisfaction with the item nor recognition of innovation.
2. The need remains unsatisfied, but the consumer or user chooses the item for various reasons.
3. The need remains unsatisfied, and nothing happens.

When perceptions exceed expectations, there are three distinctive outcomes for innovation:

1. The need is fulfilled, and the item satisfies the customer (user), who may recognize this as innovative.
2. The need is fulfilled, and the item satisfies neither the customer nor the user, so it is not recognized as innovative.
3. The need is fulfilled, and the item satisfies but does not generate recognition of innovation or additional actions, such as purchase, acquisition, or added value.

When perceptions equal expectations, there are also three distinct outcomes for innovation:

1. Status quo: Purchase or use is not given any consideration or is placed "on hold."
2. Purchase (acquisition): The product, service, or technology is purchased (selected) without regard to its innovation potential.
3. Rejection: Innovation is rejected and status quo is maintained.

Choice is the process of deciding to act on the innovation presented. Choice occurs when all individual criteria are met, resulting in use due to purchase or acquisition. Whereas satisfaction is intangible, choice is reality. What we do know is that choice plays a vital role in purchasing and acquisition decisions. Modeling, predicting, and simulating these choices are critical for understanding how individuals identify innovation. In the future, organizations will need to catalog, analyze, and understand these choices. Combined with cultural and ethnic issues, knowing your customer and their potential choices becomes critical for sustained innovation success.

The Importance of Information

Understanding consumer (user) behaviors, such as opinion, satisfaction, and choice, requires a large amount of data. Data becomes information when it is refined or filtered into a useful form or context.

Collecting data to define and understand the entire innovation experience involves

1. Profiling the customer, supplier, or user
2. Understanding how the user recognizes innovation
3. Finding the determinants of satisfaction
4. Understanding the consumer's or individual's general decision-making process

5. Knowing the business and regulatory environment
6. Having competitor (competing agencies) information
7. Accessing brand, product, and technology information

The amount of data is overwhelming. Yet, a technique such as analytics provides a mechanism for refining the data into a usable format. More organizations are choosing methods of analyzing this voluminous amount of data into opportunities for increased profits and value. "Big data" is key to the success of many organizations that require real-time information to operate the business and meet their assigned responsibilities (or core competencies).

Real-time information will become critical for a business desiring to become both nimble and responsive to customer needs. This real-time component will involve both accounting and financial data because organizations will need strict and timely financial and accounting controls to achieve just-in-time performance (value accumulation). This will spill over into the metrics of human endeavors, as human resources (HR) departments will need to provide real-time data on employee performance.

The phenomenon of real-time data will be a paradigm changer for most businesses and governments. Instant access will support accurate information, decisions, and judgments. Real-time data is becoming a reality in healthcare, economics, and security operations, to name just a few. For innovation, this real-time data will make or break the opportunity.

The large amount of information will require interpretation at the source or user level. This interpretation is something few know how to accomplish. Yet, without interpretation, the refined data could be useless. Analytics is only as good as the interpretation it portrays. Individuals interpret data and their surroundings and knowledge to form a judgment concerning whether an item is innovative. These judgments are a reflection of what each person experiences and the paradigms they develop about what innovation means to them. If the item performs better than our expectations, we can judge the item as innovative.

The Influence of Judgment

Judgment is a key attribute used with both satisfaction and choice to decide if a product, service, or technology is innovative. By this stage, the need is met or the need is subservient to some other factor. The individual must decide whether the item is innovative and whether purchasing or using the item will occur. To simplify this process, the individual must decide if the need is satisfied and then judge, based on their knowledge and experience, whether it meets their requirements (i.e., the performance they expect). Knowledge, experience, and performance are all key elements of judgment.

Knowledge

"We are what we know" is the best way to describe the modern consumer. We acquire knowledge through various means: experiences, observation, reading, conversation, the media, testing, and thinking (reasoning). For the sake of innovation, consumers (users) use the knowledge gained to judge whether the "item" will meet or exceed their needs. Therefore, a need may evolve over time. We may need something today but adjust our expectations with time and new knowledge. For sustained innovation success, needs must be monitored and catalogued given how knowledge is changing at such a rapid pace. Not only has the access to knowledge grown exponentially but the breadth and depth of knowledge have changed as well. Those who rely on knowledge can pursue information that was previously unavailable to consumers. The difficult aspect of knowledge is in filtering acquired information for our particular needs. Applying and storing what we know is filtered knowledge—dictated by our established beliefs.

Knowledge is heavily influenced by our perceptions, emotions, and preferences. For example, memorization of facts and dates for a history exam is tedious, especially if you do not like history. Individuals filter information with their preferences, opinions, and beliefs. We strive for harmony and therefore often connect disparate functions together as a way to explain what we observe, what we filter, and what we store. This is why we all observe events through "rose-colored glasses." Once we find evidence to confirm our knowledge, it is embedded as a belief. Beliefs are difficult to change or dislodge. We often need "hard" evidence (something we directly observe) to modify these beliefs. Marketers and advertising firms try to encourage product use through the five senses, slowly breaking down our belief systems.

Learning is a large part of knowledge acquisition. For innovation's sake, learning may be necessary for full functionality (it may also be a source of dissatisfaction). Many disdain innovation because of the requirement to learn something new. Innovation is center stage when the individual can experience the benefit without undue (restricted) learning. The age of jet aircraft provided cheap, fast service to anywhere in the world. Passengers need not learn much; the experience is the same with or without the learning. Yet, some want to learn and seek innovative products and technology. Learning is part of innovation and is certainly part of knowledge acquisition. People learn within a spectrum of methods and techniques. In fact, people learn in diverse ways, contributing to overall knowledge acquisition. Individuals use this diversity to solve problems, create innovations, and reduce labor and costs.

Experiences

Similar to knowledge, "we are what we have experienced." Our very actions and beliefs build on our knowledge and experience. We use our experiences to simulate future activities and encounters. Experiences are vital because they help us create "models" of our world. We use these models to assist in dealing with life encounters. We build and use models for judging products, services, and technology. Consumers (users) will always compare what they experience to what they expect. Humans form expectations based on experiences and knowledge. These expectations help form beliefs over time. Humans observe and evaluate what they expect and create models that help rationalize the world in which we live. These models are critical for those trying to influence consumer purchasing behaviors. These models help evaluate the encounters we experience into favorable, unfavorable, or neutral experiences. These models are generally situation dependent because humans encounter diverse situations, which generate needs and hence opportunities for innovation.

Experiences are critical for rational decision making. Those that lack experience usually encounter unplanned risks. These individuals must evaluate the experience with limited data collected and processed. The risk is in the decision made and the consequences of that decision. Individuals experience both positive and negative consequences that will shape behavior in the future. Too many risks add excitement and the value of the thrill—not knowing what to expect can be exhilarating. These individuals are prone to try "new" types of innovation. Yet, to some degree, all risk something when switching to a new product, process, technology, or service. Innovation can certainly embolden the risk taker, yet even risk, if rewarded, can be an experience itself. Planning for diverse experiences must be a concern for those involved with innovation; otherwise, the innovation may not appeal to a wider audience. Organizations desiring to increase the attractiveness of an innovation must be aware of knowledge, experiences, and judgments.

For a business or organization, it is critical to understand this process to remain competitive (continuing to add value) and hence profitable (cost effective). For the service organization, understanding the models that individuals build is critical for sustained service success. Although many service organizations collect customer satisfaction data, these surveys and focus groups miss an opportunity to collect knowledge and experience data, perceptions and expectations, needs analysis—all critical for predicting consumer behaviors (returns/repeat business, recommendations, suggestions for improvement, etc.). Without this information, it is difficult to construct a model of purchase/acquisition or forecast future sales/activity. Services businesses often function blindly without the data exposing these entities to unplanned/unknown market forces. Living in the information age without pertinent information is a certain recipe for disaster.

If experiences and knowledge cannot satisfy a need, individuals will seek a solution whether to choose an innovative item, find a substitute, or do nothing. The model developed to choose an innovative solution is of most concern. Finding commonality among cultural and ethnic groups will provide information on which to ensure item acceptance.

Performance

Performance is the key measurement for innovation creation and sustainability. Individuals expect that innovative items will meet or exceed performance expectations. This is true of all innovative products, services, or technology.

If not meeting performance expectations, then a judgment could be negative or neutral. Organizations expect that innovation will meet or exceed financial (performance) measures. How an item performs is key to understanding how and why an individual or organization purchased (acquired) the item. Performance describes the function of an item. It is a tangible measure in terms of how the item "works"; it is also intangible as it can involve the five senses. Experiences and knowledge are components of how to judge performance. From a tangible perspective, individuals will judge the item on how it meets a need. For example, purchasing the latest smartphone because it is innovative means that at least one product characteristic (or feature) exceeds the customer's expectations while meeting some need. Many of these characteristics are best described as intangible (color, feel, prestige, etc.). These requirements are different from product or service requirements, which are often more tangible.

Identifying characteristics that numerous individuals share is a key to success. Features such as size, weight, ease of use, sound quality, and the like are all tangible performance measures. The individual uses knowledge and experience to develop criteria for assessing these features. This information is of value to the organization. Organizations that collect and process this information certainly have a competitive advantage.

The same is true of identifying the intangibles based on the five senses. Features such as color, "feel," and pride of ownership (having the latest gadget) also contribute to performance. These all enter into the purchase decision. Failure to account for these criteria will result in an increased failure rate.

Innovation is often recognized when needs are satisfied, especially when perceptions greatly exceed expectations. It is possible that meeting one's individual needs does not meet all needs. Therefore, recognition of innovation is individualized. A multitude of opinions exists regarding its definition, its effect, and in fact its very existence. Those who generate innovations must understand this critical concept. Associating innovation with the introduction of a product, technology, or service does not make it a reality. Innovation becomes a reality when it satisfies a need. Individuals, using their knowledge, experience, and opinion judge an item innovative when that need is satisfied and the performance better than expected. Telling individuals that the item is innovative could be misleading, thus influencing the potential purchase or acquisition.

Inclusion

The focus of this chapter has been on coalescence: what components combine to describe the process of identifying innovation. The message must be clear, but the interpretation may be clouded by varying components that can actually negate the experience. The key is to develop a message that addresses the combined elements (as discussed in this chapter) of innovation, is inclusive, provides the evidence for innovativeness, and realizes that everyone will not judge or perceive the item as innovative.

Innovation is indeed an opportunity. Innovation is not static but dynamic in its appeal to customers (users). The better the innovator can understand, define, and quantify these elements, the more successful the innovator will be in providing innovative products, processes, or services.

Given the variety of individual experiences, expectations, and observations, focus on the benefits (improved performance) and how they meet the individual needs. By focusing the organization's efforts on meeting needs and excelling at performance, the innovation will appeal to a larger number of prospective acquirers. Inclusivity means bringing together those separated by one or more components. Innovation can accomplish this by its very nature: offering a better product, service, or technology that meets a previously unmet (unsatisfied) need.

Summary

This chapter verified that innovation is both a concept and a reality. Innovation has both tangible and intangible characteristics—a mixture of definitions, meanings, and intentions.

This chapter described the components that must coalesce to achieve innovation: from opinions to available information. Although innovation may begin with an idea, that idea must meet a need and ultimately provide value. We chiefly discussed the five elements of innovation (opinions, satisfaction, choices, judgments, and information) as they combine (coalesce) to define innovation.

Discussion Questions

1. Discuss how and what innovation means to you.
2. Why is the concept of coalescence so critical to understanding innovation? Be sure to include examples to describe your answer.
3. Explain Figure 2.4. You may use a product or company to explain the chart.
4. Describe what it means for innovation to be a "combination of choices."
5. Why do individuals use an integrated process to judge innovations? Could you simplify this process?
6. Would telling someone a product, process, or technology is innovative be enough for that person to purchase or acquire the item?

Assignments

1. Choose a recent product, service, or technology you consider innovative. Did the outcome of the innovation meet the need? In addition, describe the value obtained by acquiring the item.
2. Think of two like products (services) that should provide the same outcome; for instance, consider two LED 60-inch televisions: One has Wi-Fi, Internet capability, and sound-bar sound quality; the other does not have these features. Do features make a product (technology) or service innovative, or is there something else that resonates with consumers?

3. Select a process, service, or function within the government to innovate. Describe the need and identify the value.

Chapter 3

Cultural and Environmental Influences on Innovation

Introduction

When discussing the term *culture*, it conjures up a multitude of thoughts and emotions that range from ethnicity to the organizational climate experienced by federal employees each day. Moreover, the relationship between culture and innovation becomes even more complex regardless of how you define the two. Poškienė (2006) attempts to capture the relationship between innovation and culture as follows: "The relationship between culture and innovation is more complex than the research can reveal. It is characterized by many determinants that are simply too difficult to be expressed, measured or perceived. The impact of culture on creativity and innovation depends on the relationships built and on the nature of agreement" (p. 45). Given the unique qualities and values of individuals, we certainly agree it would be difficult to assess each individual's perspective in any organization or agency, but we do believe you can assess the overall culture. Although generalizations can characterize the majority of traits that identify their actions, beliefs, values, and behaviors within a culture, they are not absolute.

To help better define our position, anthropologists Atwater, Lance, Woodard, and Johnson "view culture as specific values, traditions, worldviews, behaviors, and social and political relationships shared by a group of humans" (2013, p. 7). All agencies and organizations develop one or more predominant cultures. These cultures can shift and change dramatically with age, financial condition, product or service mix, and outside influences (politics, societal priorities, etc.). The question is not what type of organizational culture is present but rather whether the culture supports innovation or has the ability to transform itself to become an innovative culture. With an eye on globalization and cross-border integration, industry-leading organizations leverage cultures that encompass a global economic condition, nationality, ethnicity, and history (political, judicial, business, available freedoms, etc.). Cultural traits, what we refer to as *influencers*, have a direct effect on innovation because these affect individuals who both innovate and use the innovation to improve efficiency and effectiveness.

For federal agencies and organizations to create a sustainable innovation culture, having team-wide understanding and appreciation of the organization's core competencies is tantamount. Further, knowing how those core competencies relate to innovation and the other organizations within the government enterprise requires visionary leaders, dedicated resources, and a firm intent to enculturate innovation into daily routines. This chapter describes and examines these influences from both global and individual perspectives and in terms of the federal government—its departments and functions as well as the businesses and nonprofits that support the federal sector.

Large-Scale Cultural Influences: Individual Perspective

Large-scale cultural influences might begin at the national level with the birth of a person and the person's ancestry, but the mosaic of cultural difference and social identity tend to define national culture. Individuals affiliate with a national identity or culture and see themselves as an extension of that group. Although cultural affiliations may not be that strong, individuals have a natural tendency to "belong" to a group. Individuals will associate with that group regardless of national identity or heritage. These national groups can exhibit many similar behavior patterns, which are central to understanding how individuals will evaluate, accept, and act on innovation. In essence, they determine whether an individual will purchase, adopt, or use the innovative product, service, or technology (Poškienė, 2006).

National Cultures

National cultures are the attributes of prevailing societies that exist among the citizens of a particular nation or ethnic group. Although not all citizens may follow or agree with the assessment, it is possible to generalize. How the generalizations are used may be open to further discussion; the intent is to develop a set of attributes useful for marketing and sales purposes. When discussing national cultures, Hofstede developed a model of five dimensions to represent the types of cultures found worldwide:

1. Power distance: access to and impact on authority
2. Individualism/collectivism: center on individual needs versus the good of the community
3. Masculinity/femininity: focused, aggressive, or creative

4. Uncertainty avoidance: stability of government, business, and society
5. Long-/short-term orientation: planning process and strategic thinking (de Mooij & Hofstede, 2010)

These dimensions are useful for generalizations but may not apply to a particular individual. Nonetheless, these provide a mechanism to evaluate national cultural characteristics. As a point of reference, Hofstede's characteristics are helpful in the areas of advertising, sales, marketing, and market research. In time, these classifications will be enhanced with the advent of real-time analytics.

Real-time analytics will provide instrumental information (data) on individuals and groups worldwide that will greatly assist companies focused on becoming innovative and help these organizations achieve their goals. Analytics is the process of filtering, categorizing, and classifying a large amount of information generated by consumers (users), hence affecting the instrumental role that knowledge management (KM) professionals play within many organizations today (Kennedy, 2014). Beyond pinpointing the individual needs irrespective of national or geographic boundaries, a common understanding of what innovation is and how it can be adopted is of instrumental importance to any agency or organization.

Ethnicities (Diversity?)

As much or more of a contributor to national culture is ethnic influence. Ethnicities have more to do with natural biological affiliations (language, skin color, heritage, etc.) than with cultural affiliation. Ethnicity is often associated with uncontrollable traits that cannot be easily disguised. It also permits individuals to make, sometimes falsely, claims about a particular ethnicity that may not be true for all participants.

Ethnicity has become a major cornerstone of diversity policies. With an emphasis on corporate social responsibility

(CSR), organizations must recognize the unique differences (and complementary similarities) of ethnic groups. Research has shown that purchasing behaviors are tied to CSR and a firm's corporate social performance (CSP) (Richardson, 2012). How ethnic groups are recognized and treated are important issues. The concept of "diversity" highlights the ethnic challenges organizations face. The major ethnicity classifications may not be sufficient for predicting exact behaviors. Therefore, further analysis of ethnic diversity beyond a simple classification is necessary. This assessment can be industry specific as most, if not all, federal agencies and organizations have passed the origin marker and advanced to gender and generational diversity (Kennedy, 2014).

When collecting demographic information, it is common to ask the respondent to select or choose a particular ethnicity. More often than not, the choices in the United States are white (sometimes referred to as nonethnic), African American, Hispanic, Asian, or other. Often, within each category are subcategories, such as the country of origin. For Asians and Hispanics, country (national) affiliation could easily overwhelm ethnicity as a major influencer. In many areas, regional or tribal affiliations are more important than country of origin. Often, locations (natural barriers) are also an issue. On a global basis, this additional ethnic information is critical for evaluative purposes.

Therefore, simple ethnic categorizations may provide limited information. Understanding ethnicity requires familiarity with the related culture, history, geography, and resources. The linkage to innovation is that certain ethnicities tend to have similar buying behaviors or traits. Certain ethnicities "value" things differently, such as Asians value the elderly for their wisdom, while Hispanics have strong social and religious ties. The difficulty is translating this to the individual. Marketers have been studying these characteristics for over 50 years (Richardson, 2012), further refining characteristics to specific demographic attributes. Yet, this information has not received the attention that is due and lacks specifics. Marketers assume that all Asians will respect their elders in the same manner, yet would this be true for Filipinos? What is known is that more information (a better characterization) is needed to truly represent these ethnicities and their purchasing behaviors.

These purchase behaviors are similar to what the individual values. When the value exceeds what is expected and a need is satisfied, innovation can occur. Characterizing ethnic traits is critical for understanding how value is created and utilized. The greater the detail, the better these ethnic characteristics will link to consumer behaviors.

Organizational Cultures

According to Berson, Oreg, and Dvir (2007), there are three major types of organizational cultures:

1. Innovative (entrepreneurial, creative, risk taking)
2. Bureaucratic (emphasis on rules, regulation, and efficiency)
3. Supportive (employee-customer centric)

Of course, generalizations of any culture exist within organizations internationally. For purposes of clarification, these three criteria do provide a basis on which to judge and determine the influence of an organization's culture on its employees, suppliers, and customers (users). Three measurement criteria help evaluate the general state of the organization:

1. Dominance: How frequently does the organization reflect this behavior?
2. Significance: How important (prevalent) is the organizational attribute?
3. History: How long has this attribute been part of the organizational DNA?

Table 3.1 Organizational Culture Assessment

Major Characteristic	Subcomponent	Dominance	Significance	History
Innovative	Entrepreneurial			
Innovative	Creative			
Innovative	Risk taking			
Bureaucratic	Playing by the rules			
Bureaucratic	Regulatory			
Bureaucratic	Efficiency focused			
Supportive	Employee focused			
Supportive	Customer/user focused			
Supportive	Cooperative			

Note: Evaluate dominance on a 3-point Likert scale: 1, minimal; 2, occasional; 3, frequent. Evaluate significance on a 3-point Likert scale: 1, not important; 2, marginal importance; 3, very important. Evaluate history on a 3-point Likert scale: 1, long history of practice over 5 years; 2, 2–5 years; 3, 0–2 years.

To assess an organization, complete Table 3.1. To obtain a balanced score, try for a sample of 30 or more individuals (in various positions within the organization). Average the scores and look for patterns in position title, age, education, and years of experience. Compare with like organizations in the same sector.

Innovation can occur in situations that are not conducive to this concept. Even bureaucratic organizations can innovate, but the flow of ideas will be less than in an organization that promotes a cooperative, supportive environment.

Type of Government (Political System) Influences

Primary influences on innovation are the government type, access to resources, and populace participation. Government influences compliment the Hofstede characteristics but focus more on the political and social realm. As governments change, so do their cultural characteristics. Cultural change is truly dynamic, and the change can vary from mild to radical. The amount (degree) of change is often due to its stability, history, years in existence, and frequency of change (government turnover). Stable governments, with a long history, generally experience little cultural change over the short term. Unstable regimes can experience a large and dramatic shift in cultural change.

There are five unique characteristic influences of government (political systems) on culture:

1. Availability of resources for basic needs (access to food, water, sanitation, education, medical services)
2. Open communications (press, social media, hospitality, entertainment)
3. Receptiveness to the populace (freedoms/responsibilities, government bureaucracy, trust)
4. Viability of infrastructure (transportation, communications, technology, currency, banking)
5. Policies, procedures, and taxes (licensing, regulations, tax burden, fees)

No government is perfect, and our emphasis is on innovation. For an innovative culture to exist, certain assumptions must be met:

1. Encouragement of innate creativity at the individual level
2. Ability to communicate these ideas
3. Receptive management (authorities)
4. Access to resources (technology, information, performance, experience)

Even a repressive regime can still encourage innovation by satisfying these criteria. Yet, for innovation to flow freely, the government must encourage creativity and readily accept those who want to participate. The best approach to understanding (evaluating) the influence of government is to assign a weight to the importance of each attribute (per government) and weight that against the ability to meet the innovative cultural criteria. These will have an impact on the choice (and decisions) an individual makes to acquire and use innovation.

Complete Table 3.2 according to impact measurement. High-impact items will contribute most to influencing the individual. Any cultural assessment will need to consider these impacts and their contributory effect regarding the expectations and perceptions of the individual.

Those items with a high impact will influence the person's choices (purchase, use, recommendations, and experiences). Positive impacts greatly enhance innovation; negative impacts inhibit or prevent innovation.

Affiliate Cultures

Affiliate cultures are cultural influences in smaller groups. Individuals joined to a group because of a set of shared values (common goals) define the term *affiliate groups*. A typical affiliate group consists of members from the same country or, for large countries, the same regions. But, it is not just countries or regions that identify these groups; they could be related to sports teams, church groups, hobbyists, military affiliation, patient groups, artists, professionals, fraternities/sororities, unionists (or, on the negative side, gangs), and more. Affiliate cultures can have a dramatic effect on behaviors, experiences, and information. At this level, trust is a major issue. Individuals trust each other when their values are similar. It is common for a person to ask an affiliate member about the member's history, experiences, and use of an innovative item. How much influence on innovation an affiliate group expends is unknown. This will become part of our ongoing research in the future. To estimate its effect, consider the affiliations possible and, depending on the product,

Table 3.2 Evaluating Government Influences on Innovation Culture

Criteria	Innate Creativity	Freedom to Communicate	Receptive Authorities	Access to Resources
Availability of resources: basic needs				
Availability of education: education				
Availability of healthcare: healthcare				
Open communications: media/social media				
Open communications: entertainment				
Receptiveness of the populace: basic needs				
Receptiveness of the populace: trust				
Viability of infrastructure: financial				
Viability of infrastructure: transportation				
Viability of infrastructure: communications				
Viability of infrastructure: taxes				
Policies/procedures: licensing				
Policies/procedures: regulations/legal				
Policies/procedures: taxes/fees				

Note: Evaluate using the following: N = negative impact; L = low impact (little or no effect); M = moderate impact; H = high impact.

service, or technology, the impact on the item. For example, if the product is a sports drink, exposure of sports fans to television advertising makes these individuals aware of the item, thus increasing the desire for such an item. Individuals rely on information and experience to judge innovation. Affiliate groups can provide this information to members, thus influencing their expectations in a positive or negative manner. Ongoing research will develop a measure of the effect of affiliate culture on purchase behaviors.

Family

Small-scale influences are often more powerful than national or ethnic cultural effects. Parents and family have a tremendous influence as they provide the moral and ethical framework that the person uses to define himself or herself. This relationship, although frequently modified, can exist into adulthood. These influences seem to be location dependent. The closer the family is, the greater the influence will be. All of us, at some time in the past, have used family members to help us with a decision. That is, we use their knowledge and experiences combined with our preferences to form an opinion or make a decision.

Families also exert pressure on the individual. Pressure to conform may change the outcome. Both negative and positive outcomes come from this exchange. Therefore, the individual may or may not recognize innovation due to overall outcomes. Research studies have demonstrated that younger people are attracted to inventions and discoveries. The effect family has on these decisions is unknown. Families will play a role in affecting behaviors.

Peers (Friends)

Similar to an affiliate, peers or friends are personal connections. People seek advice from peers and friends regarding use and purchase. Depending on the complexity component and the knowledge and experience with like products or technology, peer advice may be complementary or contradictory. A common theme is peer pressure.

Peer pressure may vary by age group, gender, social status, and so on. This is certainly a concern for younger individuals, although it is not often a concern for mature adults. The amount of peer pressure on innovation may be limited for mature adults but may be strong for younger individuals. Again, this influence needs further study and analysis.

Estimating the Effect of Culture on Innovation

It is obvious that culture affects life experiences, decision making, and judgments. Cultures that promote technical proficiency will use these experiences to seek technical solutions. In fact, a person develops a model of reality that is culturally bound (influenced). They use the model to build their knowledge and experience for judging such concepts as innovation. Knowledge of how much or how little culture influences innovation is a necessity if the organizations plan to remain competitive and distinctive. Figure 3.1 displays how these cultural

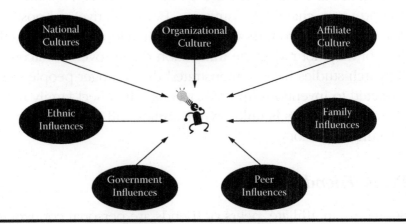

Figure 3.1   Individual influences on innovation.

influence elements relate to or affect innovation. Innovation is just one such experience as the person must decide to use or purchase the item. It is the decisions and actions that follow this evaluation that are of most interest to organizations.

Culture and the Environment

1. Make innovation a central theme in your organizational culture. This must be a top-down approach in which leaders are proponents and catalysts for an innovative culture and open idea sharing, business models, and strategies reflect and support innovation.
2. Create an open and collaborative culture and environment free from fear and in which creativity is embraced and supported by leadership. The realization that great ideas come from all rungs in the organization is essential. Not every person is an innovator, but each does have creative thoughts and ideas that could be combined with others to result in innovation. Internal and external collaboration and teaming are definitely a consideration here.
3. Change agents are everywhere (internal and external to your organization or agency). Establishing a defined process for sharing new ideas for products, services, or processes is essential. Freelancing is encouraged, but there must be a methodology for elevating ideas to the decision makers without marginalizing or changing the ideas to reflect current product offers, services, or processes.
4. Accept that change and risk taking are realities if organizations and agencies plan to grow, gain competitive advantage, and create value for their shareholders, stakeholders, and the community (McGrath, 2011). Encouraging risk at the appropriate level is essential as creative ideas and proposed innovations can diminish organizational resources without reward (Caraballo & McLaughlin, 2012).
5. Create balance. Organizational strategy must support both innovative and sustainability strategies to survive

in today's rapid-paced global market environment. Every creative idea is not adopted, and if it is, there is a good chance it will require a transition to a sustainment strategy. Forecasting the impact of launch or change, life cycle, and sustainment phases are important in assessing the return on investment.

6. Encourage an environment of trust, sincerity, diversity, human dignity, and respect for all. An environment of trust and sincerity breaks down many traditional barriers to communication. Encourage people to question the current state of play and the way things are done (Capozzi, Dye, & Howe, 2011). Diversity brings a fertile environment as differing views provide an enriched perspective while creating opportunities for expanding one's thoughts. Human dignity and respect are the bedrock that provides every player with equal footing, a sense of personal value, and the confidence to openly share thoughts and ideas.

7. Know your customer and industry. Focus on the customer and industry needs, not the wants of your organization. Creating blue ocean opportunities or white space requires a desirable product, service, or process that is a game changer, opens new market space opportunity, or improves efficiency, value, or competitive position (Kim & Mauborgne, 2006). Alliances and joint ventures are potential advantages for gaining external ideas that could provide fresh ideas.

8. Cultivate the innovation environment and mindset by providing your people with the tools, time, and training. Along with making innovation a main theme in daily operations, it is important to ensure that your most valuable resource, your people, are engaged in the innovation process, celebrate it, and understand how to package, present, and offer ideas (McCreary, 2010).

9. Challenge your organization with clear and concise expectations. When people are challenged and

expectations are known, they have an opportunity to coordinate with others, creating opportunities for real synergy (Isaksen & Ekvall, 2010).
10. Minimize conflict through collaboration and openness. Conflict creates adversity, which derails the transfer of tacit knowledge and intellectual capital development (Isaksen & Ekvall, 2010). Communication and collaboration suffer, and the benefits of teaming, creativity, cooperation, and trust are all put on hold.
11. Empowerment and accountability are important. When people feel empowered to make decisions and subsequently are accountable for the results, they actively engage with the organizational entities around them.

Use of Alliances in the Public Domain

Phelps (2010) maintains that organizations that seek partnering relationships and strategic alliances built on open collaboration and information sharing with alliance members facilitate not only access to diverse information and resources but also increased resource creation, competitive advantage, and ultimately, survival in competitive market sectors. His work is supported by a 10-year longitudinal study of alliances and relationships involving 77 information technology (IT) companies and augmented by research of nearly 40 years of published research articles.

Phelps (2010) asserts there has been more emphasis on the outcome of the alliances and little attention provided to the actual structuring of alliances, network density and composition, and the intellectual and structural capital differences between partnering entities (network-level IT diversity). These differences or "structural holes" limit the alliance's ability to access, cultivate, and share knowledge, reducing the opportunity for all players to benefit from the technological and intellectual differences of their partners. Phelps further suggests that adding more members (increasing network density) to the alliance will strengthen the cohort and positively influence exploratory innovation (creation of novel technological knowledge relative to existing knowledge stock).

Logically, the pooling of resources (intellectual and structural) in a collaborative effort could be a productive endeavor with strategic and competitive implications for alliance partners. Intellectual capital is the catalyst for competitive purpose and value. Companies are unwilling to share that information freely to remain competitive. In fact, most organizations have strictly defined limits on what they actually share. Identifying alliances is a primary objective for smaller companies to improve competitive advantage. Further, top industry leaders, such as Apple and Samsung, have forged a long-term alliance that continues to have trend-leading benefits for both firms. In this relationship, Samsung's key contribution (to the A4/A5 microprocessor) to Apple's iPhone and iPad products helped Apple redefine the cell and tablet markets and become the industry's lead innovator. Things were good in the alliance until Samsung released the Android platform in direct competition to Apple's product line—then they went to court over copyright infringement. Thus, Phelps's theory of a dense network of collaborative organizations sharing intellectual and structural capital is desirable in some, but not all, contexts.

Although some companies do not fare well by acquiring and merging to build competitive advantage (eliminating competitors and diversifying their portfolios), joint ventures and strategic alliances are possible alternatives (Porter, 1987). Hagel and Seely Brown (2005) share a similar view as Phelps (2010) regarding the importance of strong network alliances but add that these time-sensitive, collaborative approaches must be built over time through "trust ladders." These ladders of trust start with small projects and agreements with the intent to build a collaborative partnership and trust between two companies (and cultures) in hopes to reduce risk, cost, and loss of competitive advantage while building innovative approaches and products. These authors also agree that how and why alliances are formed (the rules of engagement, levels of sharing and responsibility, and defined benefits for each organization involved) are directly related to how well the alliance functions, how each member benefits, and how long that relationship lasts.

A Leader's Role in Creating a Culture of Innovation

Cultivating a positive culture for innovation starts with leadership. Regardless if it is at the individual, first-level supervisor, department, or chief executive officer level of the organization, someone leads the initiative or introduces the seed by choice or by chance. A sage piece of advice applies to this situation: "People don't fall up the hill of success." In reality, respected leaders build the mountain and rise with the mountain. In response to the question of what leaders can do to create cultures of change and innovation, we provide the following five considerations.

1. Lead by example. Emulate the behavior and conduct you expect from the innovation experience and culture. The leadership sets the tone, and subordinate leaders will most generally follow. People know when your actions do not align with verbal messages, e-mail traffic, the mission and vision statement that provides your expectations, code of conduct, business strategy, and rewards system. People know when the organization is sincere and when it practices the values of respect, appreciation, and trust. Avoid marginalization by a hierarchy of managers and supervisors who are not in line with the mission and values of the organization. Be visible, action oriented, and on point by knowing your people to the greatest extent possible. Achieving success requires that managers explain employee expectations in person, not via e-mail, announcement board, or second-party interpretation.

Be an accountable, responsible, objective, and respected mentor and coach—grow the leadership that will eventually replace you (McGrath, 2011).
2. Focus on your customer, organization, and industry (sector) needs. Consider crossover applications and opportunities related to other industry segments. Select and appoint a chief innovation officer and provide this person the resources to champion the organization's innovation process to take an idea beyond the creative thought phase (Caraballo & McLaughlin, 2012).
3. Support, foster, and enrich an organizational culture in which innovation is a core competency and strategic objective. Provide people throughout the organization with regular training, tools, and time to corroborate and collaborate openly and freely without fear of attribution. Encourage, support, and plan recurring collaborative opportunities (both internal and external) for people to share tacit and explicit information to generate new knowledge and intellectual capital. Plan and measure (what to share and when to share) events; consider spontaneous events to keep the process fresh and vibrant. Train and equip your people for success and celebrate those successes as an organization.
4. Challenge the current state of play and accept the reality that there will be failures and setbacks. The blame game is not productive and diminishes opportunities for people to share what they learned from the event and ideas on how to change the outcome the next time around—these are wonderful opportunities for continued process improvement. Capozzi, Dye, and Howe (2011) maintained that leaders should decisively question, on a regular basis, the answers and assumptions that result from challenging organizational strategy, vision, and process. Markets and cultures evolve, and the team must adapt and evolve even if you chose to follow a sustainment strategy versus a disruptive innovation strategy (Raynor, 2011).

5. See the bigger picture: the dream. You may be the one in charge and responsible for the organization and the future, but that does not always make you the resident expert or subject matter expert on all matters. This requires you to immerse yourself in all facets of the industry, your organizations, and your primary resource—your people (Capozzi et al., 2011). Be humble and honest about your capabilities and open to the input of others when it comes to sharing road maps and strategy. Dreaming lets you envision a reality that does not yet exist: Sometimes you meet someone who provides that one crucial ingredient or thought that makes it possible to leap to the next level.

Summary

Our research has provided more than 60 accepted definitions of innovation around the globe. From the repeated results (measures) and confirmation, we feel our pioneering approach in defining innovation from an individual's perceptive provides organizations involved in the pursuit of innovation with a concrete and practical definition. However, this broad range of definitions and implied ambiguity make it difficult for organizations to translate these into actionable information that supports the decision-making process. We celebrated the diversity of definitions and intents but provided a method for refining the focus. At a foundational level, refining the definition of innovation from a cultural, generational, gender, and professional point of view generates three distinct categories (new, improved, and change). This not only evolves the innovation body of knowledge but also provides a comprehensive solution that can be beneficial to the federal sector as well as the international community at the organizational level to

- reduce the confusion of what innovation is and is not;
- align team members and resources;

- better identify innovation opportunities in a much shorter time frame; and
- provide a common reference point for individual understanding of innovation to ease participation in the idea generation process.

In closing this discussion, the personal assessment is that all three approaches and their supporting principles as presented in the readings have application and appear valid for profit, nonprofit, and public-sector domains. Further, one could contend these approaches could lead to the reinvention of management and establishment of an organizational culture tempered in trust and commitment. Do we see any patterns forming throughout the theories discussed in this chapter? In our opinion, this is most certainly the case. They all appear focused on deliberate and planned process change, openness and transparency, unbridled collaboration and engagement across the organization, and most importantly, a culture and environment founded in mutual trust. In sum, the common theme is the need to change or need to reinvent the management and leadership styles in organizations to provide the same posture for success in a fast-paced and rapidly transforming business environment.

Discussion Questions

1. How would you define culture?
2. What are the cultural influences affecting you and your organization? Name five in each category (individual and organization).
3. What are the five unique characteristics of government (political systems) related to culture?
4. What are alliances and partnerships? Why are they important? As a leader in your agency or organization, what are

some of the key considerations when entering a partnership or alliance?
5. As a leader in your organization, briefly discuss three things you can do to build a culture of innovation in your organization.

Assignments

1. Considering your place of employment, complete the organizational culture assessment (Table 3.1). Briefly discuss your findings and thoughts on your results.
2. Using your organization as a study case, complete Table 3.2. Briefly summarize your results. What are your perceptions and findings?
3. For a typical organization, estimate the percentages that each cultural factor (Figure 3.1) contributes. Describe the top two influencers in more detail.

Chapter 4

Dynamic Capabilities: The Power of Innovation

Introduction

As many government employees will attest, this generation of federal-sector employees has entered unchartered waters, resulting in fear and an increasing level of uncertainty regarding what the future holds. Coupled with the effects of sequestration* (and the failed attempts by our elected officials on Capitol Hill to embrace bipartisanship and resolve their political differences), there is not only national but also international concern. With the symbiotic relationship shared by the 36 acknowledged world economies, most of these economies are becoming increasingly dependent on each other in varying degrees for continued economic growth and prosperity. As with any relationship of this nature, the evolution of important notions such as cross-border integration and globalization are changing the very fabric of the international community, a new course if America, the global champion of democracy

---

* Sequestration includes a general hiring freeze across multiple federal agencies, a targeted freeze on government employee pay over the next five years, and a significant reduction in government spending.

and capitalism, stumbles or falters. In the midst of uncertainty, government agencies have difficulty in accomplishing the tasks set before them with fewer resources and personnel than are traditionally accustomed.

In reality, environmental turbulence is only one of many dynamics driving both private- and public-sector entities to make a concerted effort to identify, understand, and pursue dynamic capabilities to adapt to these uncertain times and the fluid nature of the international landscape. Often attributed to the introduction of new technologies, cross-border integration, and the ever-increasing changes driven by international influence and globalization, some federal agencies have simply found themselves at a standstill, contemplating what this all means. The government is beyond the initial shock and awe that a smaller government and reduced discretionary spending are realities. It is time for a new type of government leader to take the stage—a leader with the hallmarks of courage, integrity, respect, knowledge, stamina, and the admired ability to build teams even in the worst circumstances. These leaders are not looking for a "silver lining" or making cases for why their missions will fail because of sequestration. Government leaders must do more with less; they lead by example and are on point. Leaders are not quibbling or making excuses; they are busy preparing their organizations for an inevitable culture change that will test even their most experienced employees. In essence, they have already enlisted the focused attention of their team to find new, efficient ways to take the crate of lemons and sell lemonade or perhaps a new product or service. The leadership mindset, paradigm, and culture must shift from one of survival to one of opportunity.

What are dynamic capabilities, and why is this concept so important? In a broader context, dynamic capabilities are routines (procedures or processes) performed at varying levels within the agency or organization. At the organizational level, such a capability is a high-level routine (or collection of routines) that, together with implementing input flows (inputs and outputs from other processes), provides the agency's decision makers (leadership and management) with an objective view or set of decision options (actionable information) for producing significant outputs to meet a particular objective or goal. The key here is objectivity: fact-based actionable information free from personal feelings and motives. Further, emphasis should be placed on the idea of *routine*—behavior that is learned, highly patterned, repetitious or quasi-repetitious, founded in part in tacit knowledge—and the specificity of objectives (Zollo & Winter, 2002). When these routines are process mapped and shared across the agency, boundaries are established with an idea of open and transparent communication.

Dynamic Capabilities

Dynamic capabilities are those skills needed to build internal and external competencies within an integrated workplace conducive to a changing environment. In other words, these are skills that determine an agency's or organization's ability to "integrate, build, and reconfigure internal and external competencies to address rapidly changing environments" (Teece, Pisano, & Shuen, 1997, p. 516). The idea and recipe for pursuing dynamic capabilities is industry and sector agnostic. For federal agencies, developing dynamic capabilities might start with experience accumulation, knowledge articulation, and knowledge codification in a deep dive of their operational routines. Within any given government agency there are traditional functions, one of which is continuing education and training. For example, learning competencies associated with these efforts are often outlined and guided by regulation or directive to facilitate a deliberate and effective learning process. Dynamic capabilities supporting the learning process compliment and support the organization's culture, core competencies, and strategy. The resulting tacit-to-explicit, explicit-to-tacit knowledge exchange can generate new ideas codified for explicit use. This provides an incredible opportunity to harvest tacit knowledge within the agency that is often difficult to access when employees are reluctant to share their ideas and thoughts for fear of ridicule, of appearing less knowledgeable in front of their peers, or simply insecurity that their ideas are important to the agency. Sometimes termed as latent or hidden dynamic capabilities, learning environments show tremendous promise for building trust ladders between leadership and employees.

Competencies are not essentially physical matters; they manifest themselves in ways such as human capital (leadership), structural capital, processes, and strategies that build the very fabric that makes a company what it is and what it can grow to be—without them, an agency's destiny can turn toward incoherence and a lack of focus. Developing leaders to function in a demanding, dynamic, and fast-paced environment is paramount.

Dynamic Capabilities and the Leader

Integrating, building, and reconfiguring internal and external competencies to address rapidly changing, high-velocity environments influencing any organization clearly present a challenge to management and leadership. In fact, the most important dynamic capabilities of any organization are their leadership, strategy, and culture. Agency leaders can employ three avenues in preparing and adapting their culture and capital structure to be more responsive to high-velocity volatile environments:

- Sound and experienced leadership (concern for the human capital, engagement for taking advantage of opportunities when they arise, result-oriented mindset, etc.)
- Ability to recognize and effectively manage change

- Engagement with dynamic capabilities by acting as a catalyst

Leadership coherence is paramount—the coherence is focused on proven and dynamic capabilities and competencies to provide leaders across the agency the opportunity to focus on congruent agency strategies that build on and fuel growth, innovation, flexibility, and resilience. In reality, when an agency or organization deviates from the things it does best (core competencies) for the risk and rewards, it ventures into uncharted waters and risks running aground. Dynamic capabilities that support an agency's core competencies are built brick by brick and form the foundation of the agency's strategic framework (architecture). The benefits of having a working knowledge of your dynamic capability road maps are significant and directly correlate to how effective and efficient an agency operates both internally and externally with shareholders, stakeholders, and its customer base. The resulting road maps serve as references or guides for agency personnel when faced with turbulent and volatile environments. Mapping of dynamic capabilities can also build leadership competency, support agency succession plans, and contribute to future strategic planning, programming, and capital investment. In an article for *Organization Science*, Martin (2011) theorized that executive leadership groups and dynamic managerial abilities act to perform an instrumental role in preparing and adapting an agency to capture and identify innovation opportunities. As agencies move through preparing and adapting their agency climate, opportunities develop or surface by collectively sensing, seizing, and reconfiguring resources to exploit those benefits.

Much like the success of private industries with open and transparent communications up, down, and laterally within a company, federal agencies have the distinct opportunity to adopt, mimic, or modify these practices to meet their specific needs. Tantamount to adopting this approach is creating a culture and environment where creativity is celebrated, employees feel valued for their contributions, and rewards exist for those who bring value to the agency. Unfortunately, chiefly due to policy and law, most federal agencies simply do not recognize federal servants for innovative ideas. In reality, generational differences have led younger, more dynamic thinkers away from public service because of its aging industrial-age mentality, practices, and culture.

There are distinct benefits for choosing a career with the government. Serving the nation was once considered a respected profession. Government service often conjures up a regimented and time-driven approach in advancement, stagnancy in the workforce, a general reluctance to change and innovate, and an established homogeneous talent pool that has led to a decade-old hangover from an industrial age mentality. In comparing the public and private sectors, dynamic capabilities and leadership in the private sector are often viewed as essential core competencies and prolific hallmarks of great companies such as Pixar, 3M, and Boeing. These companies also have a flexible but determined approach in building human capital that is responsive and agile to innovation and change-driven environments.

A Case for Process Mapping

An acknowledged tool associated with Six Sigma and value analysis, process mapping is often overlooked (or underused) by many federal agencies because of the time investment required for applying this tool correctly. Unlike many private-sector companies, the federal government is known for its affinity to detail-oriented standard operating procedures and processes—from the top down. In reality, this affinity for documentation is one of the prominent reasons the government enterprise is labeled as Napoleonic, bureaucratic, and mind-numbingly inflexible or incapable of adapting to environmental changes in a timely manner. At the enterprise and organizational level, it can be a hit-or-miss proposition whether processes add value or competitive advantage to the mission or core competencies. One important tool that many agencies overlook, due in part to higher priority tasks and the time it takes to complete, is process mapping. Further, agencies often do not commit the necessary time to perform process mapping correctly, and inevitably do not devote the correct resources that understand how it contributes to accomplishing the organization's objectives, mission, and vision.

One method used to view how a process functions is process mapping. Process maps are often used to display best practices. Process maps can lead to the creation of best practices. There is no doubt that federal agencies review their standard operating procedures and processes on a recurring basis, but the question remains whether those reviews are accomplished by knowledgeable personnel who have a firm grasp on how the scaffolding fits together to accomplish the mission.

Managers often assign employees or interns the task of creating process maps. These individuals lack the necessary background or knowledge to recognize value-added or non-value-added steps or tasks. Mapping the organization's critical dynamic capabilities or routine tasks that contribute directly to accomplishing an agency's core competencies is not a task for an intern or just anyone who has time to complete the task. Experienced and knowledgeable staff members should lead these efforts with a critical eye for detail and the ability to analyze each step (task) for its influence on primary goals. Understanding how these tasks interrelate and how these tasks contribute to the chain of associated tasks will provide valuable information. Granted, not every task an agency performs is critical, and lesser routine tasks are excellent opportunities for inexperienced personnel to familiarize themselves with the existing process. Further, a fresh set of eyes on tasks that have been reviewed by more seasoned personnel can be beneficial as new questions can be introduced that offer a fresh perspective.

Although the roles and responsibilities of mid- and senior-level leadership within an agency have competing priorities for their time, they must be involved in the process because the results ideally should ultimately encompass best practices. Distractions take away the focus and intent of the exercise. Many leaders move from meeting to meeting throughout their day. Process maps provide a simple and effective means of sequentially displaying a process in a simple (logical) graphical form.

By no means a complete or prioritized list, we offer some immediate thoughts to consider when selecting a leader and team members for mapping critical processes:

- Knowledge of the agency's core competencies; relationship with the dynamic capability, practices, procedures, or process (does the individual or team of individuals perform this task and how often?)
- Experience and attention to detail (look for staff members with a displayed history of continuous process improvement)
- Creative and innovative approaches
- Opportunity cost of investing in mapping a given process or set of processes

Benefits of Process Mapping for Innovation

Visualization of images and pictures has been used by virtually every recognized civilization known to humankind as a primary means of communication and transcending language barriers. From modern sports playbooks, to the Greek language spanning some 34 centuries of written records, or to the 750 picture signs or hieroglyphics ancient Egyptians used to communicate, record, and capture events, effective and efficient communication is tantamount to both public- and private-sector organizations. In a manner of speaking, process mapping is as much a language as it is a visual comprising its own strategic, operational, and tactical implications. The old adage that "a picture is worth a thousand words" seems to summarize best the significant roles process mapping plays in the increasingly diverse, competitive, and dynamic global landscape of business and politics. By design, process mapping combines the written word with graphs, pictures, and images to add clarity in communicating complex or process-centric activities. In reality, a well-constructed process map can have game-changing implications for federal agencies. Although the following is by no means a complete list of potential applications, process mapping can help federal agencies seeking innovation opportunities to:

- Add value in processes by isolating value-added steps in a process, identifying problems in processes, or identifying waste attributed to non-value-added or outdated processes
- Compliment the written process while providing visualization support for communicating difficult or complex ideas (essential for training plans, knowledge sharing, etc.) and compliment modes of learning (online, traditional brick and mortar, or self-paced)
- Contribute to the ease and speed for analysis and identification of potential efficiency and incremental innovation opportunities (where processes can be improved/changed) and identify root causes of poor performance and productivity
- Establish the foundation and baseline for validating and verifying requirements and justifying capital expenditures
- Identify capability gaps in core competencies and definition of requirements as needed by the Planning, Programming, Budget, and Execution (PPBE) process for capital investment decisions
- Indicate compliance with regulatory guidance and protect and preserve intellectual property and business intelligence

- Reduce operating costs, risk, and waste while increasing productivity, improving understanding, simplifying complex processes, and improving outcome or experience for customers
- Position the agency to effectively manage its dynamic capabilities

In reality, many federal agencies and organizations view process mapping as a monotonous and time-consuming practice with little value. It does require a disciplined and objective approach and support from leadership. Process mapping requires documentation of procedure, practice, and process. This tool provides value in aligning an agency's activities with its core competencies. When faced with outdated practices or processes, mapping those processes delivers an objective evaluation essential for discovering non-value-added steps, potential opportunities for improvement (incremental innovation), and the adoption of new practices.

Lean Concepts

Often attributed to the actions Toyota initiated in the 1950s to enhance their production system, the concept and principles of Lean have not only transformed the ways businesses develop and grow but also have influenced key strategy documents (e.g., business models and strategies) that drive an organization's or agency's very existence. Adoption of the term *Lean* did not occur until the 1990s; the ultimate focus of Lean is reducing waste in the value stream through efficient and effective processes. As the federal ecosystem has embraced Lean by adopting principled and disciplined approaches to meet their goals and objectives, the N²OVATE™ methodology provides a decision process that is focused, disciplined, efficient, and effective in reducing waste, adding value, and

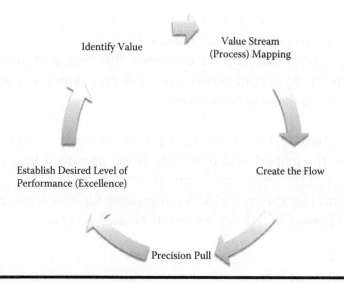

Figure 4.1 The Lean principle cycle.

improving the successful outcome of an innovation decision. In relation to the federal sector, Lean has become an expectation and a common term in the federal sector lexicon across several organizations and agencies.

A typical five-step Lean thought process for guiding Lean techniques is presented in Figure 4.1. This in no way indicates the process is straightforward or easy to implement in any market segment or industry; it simply provides the example touch points and steps of Lean practitioners and how they evolve into transformational methodology.

John Shook's (2014) Lean transformation model offers five distinct questions in creating value and the importance of aligning three key factors (process, people, and purpose) in the pursuit of continuous innovation:

1. Purpose: What is the purpose of a change?
2. Improvement: If we implement a change, how are we improving the actual work?
3. Capability: What (how) are we building (value-added) capability?

4. Champion: What does management and leadership need to do to champion the change (new way)?
5. Culture: What basic assumptions, thinking, and mindset are likely to compromise the existing culture, and who is driving the transformation?

In summary, having the right people in place, being deliberate and disciplined, and ultimately being grounded by a common set of principles are essential not only to a proven concept like Lean but also in the decision process for how innovations are addressed in federal organizations and agencies.

Cause and Effect

There is a diverse set of thoughts across the government enterprise on the value that Six Sigma and its associated tools brought to resolving what is widely acknowledged as a Napoleonic, industrial age bureaucracy mired in complex processes that thwart innovation. In reality, several organizations struggled with the simplest practice of individuals within their own community asking "Why?" until the answers to those whys stopped and someone retreated from the conversation with "That's just the way it is" or worse, if in a position of power, directed or ordered the conversation to end. Yes, we do agree that Six Sigma was not the answer to every question, problem statement, or need the federal government entities faced, but not using the tools and trained evangelists to contribute to answering difficult questions is a sore spot for many who still believe there is a use and place for these tools and the approach. It all boils down to knowing the right tools, at the right time, and for the right reason. This is simply a basic tool that can unify purpose and build common understanding of a problem or requirement. The tool helps identify the root causes of the problem and certainly culls the potential root causes that appear to be a major hurdle for many leaders to identify, accept, and act on if they so desire. In the case of cause and effect, it is a well-known relationship and rule we all generally acknowledge and experience in our day-to-day lives. Hence, we feel the process is adaptable, and we consider it "leadership-level" agnostic because it can be a personal (individual) or team-level experience based on the desired outcome.

Typically, there are four steps in creating a cause-and-effect diagram (also known as a fishbone diagram or an Ishikawa diagram, named after its creator). Please note that all four activities identified should be approached in an objective manner (bias has no place because it leads to proposing solutions and preconceived ideas that often can be key contributors to why you have a problem or need in the first place). That said, this does not mean you vacate the experience of the participants. All participants enter as equals and all must bring with them a disciplined and open-minded approach (nonattribution environment). If you are a senior leader and do not like the realities or the thoughts of those who work your assigned missions on a day-to-day basis, this approach is probably not for you. As a point of reference, Figure 4.2 displays a typical cause-and-effect diagram.

1. Define the problem (requirement or need) you want to solve or explore.
2. Identify potential causes that are key drivers of the problem, requirement, or need.
3. Explore and capture the potential reason(s) behind the potential cause(s) you have identified.
4. Seek to capture (identify) the reasons behind the key causes you have identified.

Begin by identifying a problem; write down the major categories (the bones); provide each participant ample time to brainstorm ideas (reasons for the problem). Add ideas to each category and have each individual add to the diagram

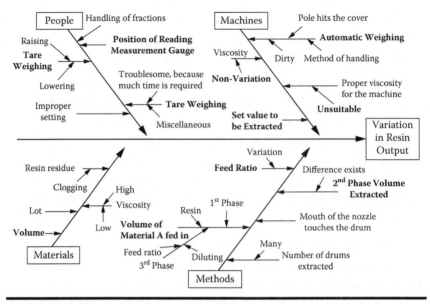

Figure 4.2 The cause-and-effect (Ishikawa) Diagram. (From Clark, Gordon M. Statistical thinking to improve quality. March 28, 2008. http://www4.asq.org/blogs/statistics/2008/03/. Accessed January 15, 2015.)

and look for integrated reasons or causes for the problem. Continue the process until all ideas are exhausted.

The benefit of the cause-and-effect tool and approach is that it provides a simple and basic tool for synchronizing and focusing participant ideas and thoughts on the defined problems, needs, and their potential causation. Again, it can be tailored to a specific problem statement, requirement, or need by simply changing the key elements in the diagram. For instance, some examples of key contributor (category) approaches follow:

- Suppliers, surroundings, systems, and skills
- Material, methodology, manpower, machines, and environment
- Product, price, people, promotion, physical environment, productivity, place, and process

The typical key contributor categories employed depend on the industry. In the manufacturing and production industries, machines, people, methods, materials, measurement, and environment are considered. In service industries, policies, procedures, plant/technology, and people are key contributors. If process steps are the target set of your organization or agency, you might consider items such as policies/procedures, measures, and resources. By no means is this a complete list; these are simply ideas for consideration. The cause-and-effect or fishbone diagram provides only the causes; it does not rank or rate each cause. Apply a rating scale (e.g., 10/5/1) to causes to determine which cause has the greatest influence on the problem. Each person gets 10 total votes; more than 5 votes cannot be assigned to a single cause. The minimum that can be assigned is 1 vote. If it is a complex problem, expect more than one cause or a set of interrelated causes.

There are some things to avoid when employing a cause-and-effect diagram approach. One is certainly the "free-for-all" mentality and knowing when you are beyond your span of influence and control in making changes to address the problem, requirement, or need. Engaging the support of an experienced facilitator helps ensure the team remains focused on the problem statement or requirement (need). It is also prudent to ensure the rules of engagement (ROE) are defined and accepted before the team starts the process. Be open to team members who do not share the same view as you unless you want to lead the result to your way of thinking (this practice is not advisable). Further, if possible, add members to your resolution team who are familiar with the topic and have ownership responsibility at some level of the problem or requirement (need). Again, opinions may be more belief than experiences and knowledge. In reality, narrowing the scope of your problem statement and avoiding a wide range of endless possibilities for a cause can be a monolithic task. The ultimate goal should be generating new knowledge in an open and transparent environment while identifying opportunities for the decision maker to consider. We have experienced firsthand those events that lead teams in building well-scripted recommendations to a decision maker outside their circle of influence. If you approach this tool with the idea that you are not concerned with who receives the credit and you just want the issue resolved, you are a perfect candidate for a cause-and-effect project within your organization or agency.

Summary

By necessity, agencies must move forward and seek new ways to adapt their organizational culture and dynamic capabilities (processes) to a resource-disparate environment where the demand for changing needs is constant. Agencies that do this well have a much better chance of emerging from this time of opportunity with an agile, responsive, and tempered culture built to withstand virtually any environmental uncertainty. This recent trend has focused not only on the impact dynamic capabilities have on the organization in changing environments but also on the implications that dynamic capabilities may have for an organization's strategy development and leadership.

In summary, leaders and managers should make every effort to concentrate on building learning environments within their organizations, dynamic capabilities that compliment or function like core competencies. This occurs while focusing leadership, financial capital, staff relationships, reputation, technological know-how, and strategic thinking, which all contribute to the overarching and long-term strategy of building a flexible, enduring, learning, and innovative organizational culture. Further, we provided a general definition of what dynamic capabilities are, potential ways to develop those capabilities (to include competencies), and how those capabilities may influence the leadership growth approach.

Discussion Questions

1. What are dynamic capabilities? Cite three within your company, agency, or organization.
2. Based on the discussion in this chapter on dynamic capabilities and using your own research, develop a position or case for why (or why not) your company, agency, or organization might struggle or easily adapt to rapidly changing environments.
3. What is a competency, and what is a process? How are these related? Why is understanding a process important in determining value in company, agency, and organization competencies?
4. What are process maps, and why would they be useful in your company, agency, or organization?
5. Using the discussion points in this chapter and your own research, cite three potential reasons why government agencies do not use process mapping as a key tool for identifying value in their processes.
6. What is a dynamic capability? Use an example from your own experience to clarify understanding.
7. What role does leadership play in developing dynamic capabilities?
8. What is process mapping? Identify the benefits achieved through process mapping.
9. What are some examples of Lean concepts? Pick three examples where they would apply in your organization or agency.
10. What is cause and effect? Recommend four key contributing categories for a cause-and-effect session within your organization.

Assignments

1. Consider your own company in light of the dynamic capability discussion. Take one department or functional area within your company and identify three key process functions or competencies. Using the process map approach, build process map flows for the three key processes and build a short discussion of each of these steps in relation to the value they bring to each process. Complete a final assessment (in your own words) with potential recommendations for improving each of the three processes if there are opportunities for change and improvement.
2. Identify and research a company in the private sector or an organization or agency in the public sector. (Hint: your own company or industry will speed the research and needs identification phase but be careful of your own bias and of jumping to conclusions through assumptions.) Objectively identify and assess one to three critical requirements (needs) you feel are germane to your selected company's or organization's observed requirements. When you have your requirements (needs) identified, objectify (develop objects) for each of those needs and provide a brief discussion to capture your results.

Chapter 5

Needs and Requirements: How to Initiate Sustained Successful Innovation

Introduction

Innovation begins by defining needs (requirements). For discussion purposes and in general, the term *requirement* is synonymous with our definition of *need*. In some manner, needs are related to wants in the same sense that many federal organizations understand there is a need and desire for change but struggle with implementing that change, thereby leaving the change unfulfilled. Subsequently, requirements and needs are synonymous. This is not to say that wants (e.g., enhancing capabilities) will not eventually become needs based on changes in policy, economic conditions, or international pressure, for example. In reality, needs are often accompanied by wants as "add-ons" as people close to the decision maker seek to enhance the perceived value of meeting a prioritized need. By nature, we all tend to seek additional value based on our own personal experience, but the subjectivity this activity brings often changes how we accomplish our goal of satisfying a defined need. This chapter shares our thoughts on the definition of requirements as they relate to core competencies. It also briefly introduces the Defense Acquisition System (DAS), Joint Capabilities and Integration Development System (JCIDS), and the Planning, Programming, Budget, and Execution (PPBE) process relationship. Finally, in this chapter, the terms *agency* and *organization* are synonymous, as are the terms *requirement* and *need*.

Requirements, Core Competency, and Mission Set Definition

Although we accept the idea that wants are often associated with a need, defining and prioritizing needs in an effective, efficient, and timely manner are the main challenges government leaders face. Definition of requirements (need) is a time-consuming and complicated endeavor by nature. Stripping out the wants is essential in properly defining, understanding, and clarifying what a requirement (need) is, and it becomes a primary reason why organizations struggle with prioritizing their most important needs. We acknowledge the challenge that satisfying a broad range of competing interests (requirements or needs) associated with organizational core competencies is not a simple task. At a basic level, core competencies are prioritized unique abilities (capabilities) or competitive advantages within a particular organization not easily imitated by other organizations. Although unique, organizations can share competencies. For the purposes of this book, mission sets are defined functions and activities focused on specific objectives an agency or organization is assigned by itself or in combination with other organizations to achieve a perceived outcome.

Beyond our own bias, habits, and acclimation toward change, there are cultural, systemic, and organizational roadblocks that prevent the adoption of the innovation mindset when approaching requirements and needs. Some of these are as follows:

- A common understanding of why individuals do what they do
- How to examine certain aspects of the innovation
- How each person defines or views the impact of innovation on their own lives

Agencies adopted the term *standard operating procedures* (SOPs) or accepted "rules of the road" to help improve process understanding, synchronization of activities, and the development of commonality in understanding. SOPs describe how to accomplish certain activities within (internal) and outside (external) federal organizations or agencies. These prescribed activities (processes) were not only designed as a unifying directive for how to accomplish goals within an agency, but also typically to indicate what is perceived as the best practice for accomplishing things in the most efficient and effective manner. Notice that here reference to the word *timely*, often a key performance parameter in project or program management activities, is missing. We feel this is simply one of the major shortcomings in the bureaucratic federal complex that requires immediate attention.

Unfortunately, the standard approach to scripting processes under the flag of efficiency and effectiveness can also be a hurdle for both private- and public-sector organizations in evolving their culture to one that is agile, responsive, and readily adaptive to new ideas and innovation. In reality, what one leader may think is "innovative" another may think is not based on experience or a lack of understanding. Hence, focusing on objectively defining a need or requirement is simply tantamount to pursuing any innovation. In regard to organizational culture and the propensity of an agency or organization to "exist," grow, and sustain an innovative culture, requires a full understanding of the requirements or needs of their organization. Again, try not to confuse needs (requirements) with specific desires or wants. Doing so can ultimately lead to skewing one's focus on the resource decision process when considering innovation opportunities or value propositions.

Truly, there are additional factors at play within an organization or agency that affect the adoption of an innovation culture. Some are internal (organizational approach to innovation, workforce dynamics, past practice, SOPs, etc.), others are external (e.g., the impact of sequestration, market climate or conditions, etc.). This chapter provides a few examples and thoughts on why an open and objective definition of requirements (needs) is essential in the pursuit and adoption of an innovation culture. These requirements include some internal and external factors that are intuitive to most federal decision makers but often are overlooked or discounted. Finally, a disciplined and validated approach like N²OVATE™ can assist federal organizations in achieving a track record of success in their pursuit of innovation.

One final comment is important here: Once a requirement or need is defined and prioritized, it will compete with other requirements across the organization for support and funding to become an accepted program. To have a program, you must have funding. This is where you often find that enhancements (wants) to the base requirement graduate to the "requirement" status. For example, a request for proposal (RFP) is a solicitation to outside organizations to bid on providing a product or service. Central to the RFP is a definition of what your organization's preliminary requirements are and what they are looking to procure or acquire from potential service providers. As you receive responses to the RFP, requirements can begin to take further definition based on private industry responses and capability to meet all or some of your defined requirements. As you review these submissions from service providers, consider how they add both identified and unidentified wants to increase value to the initial requirements. Going from the simple to the very complex, these wants can become instrumental deliverables in a statement of work (SOW) and subsequently a requirement or deliverable. When finalizing the contract, these are usually binding requirements or performance parameters (contract line item numbers, CLINs) the selected contractor must meet to receive full payment for services rendered. There is more granularity to this well-defined set of processes for those well versed in DAS; this is strictly a basic introduction for those who are not familiar with the process.

Acquiring Capabilities and Addressing Capability Gaps

Based on the complexity, size, geography, and functional orientation of the US government ecosystem, a brief introduction to the environment where the N²OVATE™ methodology and process applies is worth mentioning. Largely criticized by a litany of key players (within and outside) of DAS, the system is typically associated with the action phase of the PPBE portion of the strategic decision process. Specific criticisms and concerns vary but most commonly are based on the following and essentially related to the time it takes to accomplished things (achieve an effective outcome that meets defined requirements or needs):

1. After receiving approval, introduce, justify, and gain support for an innovation or program. For example, Homeland Security's answer to controlling illegal immigration on our southern border with Mexico employs the use of unmanned aerial vehicles, satellite reconnaissance, and imagery technologies.
2. There is excessive time to get the product or service to the customer/client (e.g., the world's premier fixed-wing fighter aircraft, the F-22, took over 20 years to design, develop, and deploy to the field).

3. There is a realization that the perceived value of the opportunity or value the innovation once offered might be overcome by events and the program or project cancelled (e.g., the replacement program for the US president's rotary fleet or helicopter replacement program, which was cancelled after expenditure of $2.3 billion to develop and field the solution).

For those experienced acquisition experts intimately familiar with acquisition processes and program management, there are no great epiphanies or compounded panaceas offered here. The discussion is basic and provides a general understanding of the acquisition process to someone unfamiliar with the process. Key phases involve the planning, implementation, and execution of an innovation and the inherent complexities or potential roadblocks in acquiring things by most government agencies. Additional fidelity and proposed prescriptions are topics for an additional book. To illustrate the complexities and challenges federal government entities have in the timely adoption of innovation, we offer a high-level view and example of the three major systems that have an impact on and limit the ability of the Department of Defense (DoD) to adopt a more Lean and agile approach to pursuing innovation. Chiefly speaking, the complex relationship between the DAS, JCIDS, and PPBE process is simplified in Figure 5.1.

The Defense Acquisition System

In its current approach, the DAS has five acknowledged or major phases: Material Solution Analysis (MSA), Technology Development (TD), Engineering and Manufacturing Development (EMD), Production and Deployment (P&D), and Operations and Support (O&S). Although the O&S phase is not considered a major phase in acquisition, it is a quintessential and critical driver of the requirements' definition in and of itself.

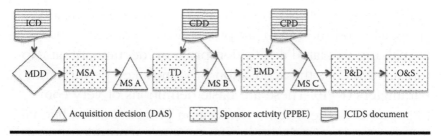

Figure 5.1 The DAS, JCIDS, and PPBE relationship.

Working in concert with JCIDS and the PPBE, DAS provides a consistent decision support process for delivering cost-effective and timely solutions to validated and verified capability gaps. Beyond the immediate fit in the MSA phase of DAS, we see multiple complimentary opportunities throughout this portion of DAS for a disciplined, timely, and customizable solution set like the N²OVATE™ methodology, tools, and process.

Managed by the under secretary of defense (USD) for acquisition, technology, and logistics (AT&L), the DAS is the principal process for transforming validated needs (requirements) into materiél capability solutions. The DAS uses the JCIDS process documents (e.g., the Initial Capabilities Document, Capabilities Development Document, and Capabilities Production Document) to provide the critical link between validated capability requirements (needs) and the acquisition of materiél capability solutions through five major phases of the DAS: MSA, TD, EMD, P&D, and O&S. Figure 5.1 depicts the relationship between the acquisition process and JCIDS documents.

The material development decision (MDD) is a mandatory step that signifies the formal entry point in any DAU acquisition process. Typically, an Initial Capabilities Document (ICD) and Analysis of Alternatives (AoA) are chief documents required to support the MDD step. A milestone (MS) is a point in the acquisition process where a formal decision is made to move on to the next step in the acquisition process or stop the acquisition effort. The capability development document (CDD) is a formal acquisition program document that supports what is termed as an evolutionary acquisition approach; and the capability production document (CPD) is a formal and validated acquisition program document that captures and refines the acquisition program's key performance parameters (KPP). These are basic definitions of very complex points in the defense acquisition process and additional information can be found on the Defense Acquisition Portal at https://dap.dau.mil.

Joint Capabilities and Integration Development System

The Joint Capabilities and Integration Development System (JCIDS) is a capabilities-based approach to requirements (needs) generation and works in parallel with the DAS and PPBE. JCIDS is the process of the chair of the Joint Chiefs of Staff (CJCS) for identifying, assessing, and prioritizing joint military capability requirements (needs). Each defense agency's requirements and acquisition process is shaped by its culture and the government scaffolding of policy and law. In general, the Air Force is chiefly concerned with the air and aerospace domain, the Army the terrestrial domain, and the Navy and the Coast Guard, the littoral zones and the nautical latitudes of roughly 70 percent of Earth's surface. The Marines simply have an affinity to operate in any of these environments. The more complex these moving pieces become, each change sets in motion the potential for some type of gap between processes associated with the change.

The JCIDS process exists to support the Joint Requirements Oversight Council (JROC) and CJCS responsibilities for identifying, assessing, validating, and prioritizing joint military capability requirements (needs). Although not a seamless or agile process, JCIDS provides some transparency in the process, allowing the DoD and its action arm (JROC) to stabilize the competing needs of the joint military environment.

By design, JCIDS provides senior federal leadership with an accepted set of processes focused on synchronizing and prioritizing the competing requirements (needs) of organizations assigned under the DoD. The system is also designed to work with participating partners (those who have equity in the capability requirements process) and provide actionable information to the right decision maker, at the right level, and at the right time.

JCIDS employs a Joint Capability Areas (JCA) approach of functionally organizing capability requirements (needs) in portfolios relative to each service's core competencies and assigned missions (responsibilities). Functional Capability Boards (FCBs) manage the resulting capability requirements portfolios in each of the service organizations. In the joint commands, the Joint Capabilities Boards (JCBs), made up of representatives from service-specific FCBs, manage joint capability portfolios for which services have complimentary interests or share core competencies. This approach allows the FCBs to find similar DoD capabilities that, by functional category, support capability portfolio management, capabilities-based force development, strategic and operational planning, capability analysis, and ultimately investment decision (the DAS website can be accessed at https://dap.dau.mil).

Planning, Programming, Budget, and Execution Process

The Planning, Programming, Budget and Execution (PPBE) process is the DoD's primary vehicle for managing resources. It is also instrumental to the procurement and research, development, technology, and engineering (RDT&E) functions typically associated with innovation within the military organizations. Similarly, noncombatant agencies outside the DoD are subject to similar, if not the same, requirements driven by PPBE. There are four phases associated with the PPBE:

planning, programming, budget, and execution. Each federal organization, like the military services, competes for resources under the national budget submission process that drew breath from the Budget and Accounting Act of 1921.

There are many key players in the US government involved in this cyclical and recurring process each year. Primary players include federal organization leadership, the legislative and executive branches of the government, and most notably, the president of the United States (POTUS). The PPBE is the process in which federal agencies introduce their priorities for funding, justify those priorities, and secure funding and resources for those priorities and operating capital to complete their assigned functional duties and responsibilities. In the case of the DoD, the Future Years Defense Program (FYDP) is the official compendium or resource database outlining what each service requires to perform its assigned missions. Updated twice a year, the FYDP is the approved budget allocated by our nation's leadership and, once approved by the president, represents each service's total obligation authority (TOA) (sum of all budget authority granted or requested from Congress) for a given year.

In the case of military organizations, this becomes a challenge in trying to pursue some innovation opportunities as they are competing for limited resources in a resource-disparate environment. Further, these organizations must deal with a regimented and scripted process with established windows of opportunity for introducing new ideas and adequately defining, justifying, and securing the appropriate level of funding and support. Within the US government bureaucracy, terms such as *agile*, *flexible*, or *responsive* rarely characterize the organization.

Managing the Complexities of the DAS, JCIDS, and PPBE Relationship

Intended to work in tandem, the DAS-JCIDS-PPBE relationship is complex and focused on providing actionable information for senior DoD decision makers. The roles each system play in this fusion of needs is unifying, with each having its own roles, responsibilities, and expectations (RRE). Fundamentally, JCIDS provides a system for capturing and evaluating potential capability requirements (needs) and nonmateriél solutions; the DAS provides a venue for acquiring materiél solutions; and the PPBE provides the resources for executing a materiél or nonmaterial solution to capability gaps and requirements. As previously provided, these systems must work in concert to provide all levels of decision makers with a consistent and disciplined decision-making process for delivering value-added, timely, and cost-effective capability solutions.

Ultimately, the trio of processes arms the DoD with a formidable construct for determining, validating, prioritizing, and initially assessing the value each capability requirement (need) adds to the owning organization's core competency set. The relationship also supports and solicits common ground among the different organizational cultures by identifying capability gaps, risks, and funding issues while developing and delivering value-added nonmateriél and materiél capability solutions.

Sequestration

Because the government has squandered trillions of taxpayer dollars on a plethora of questionable defense, social, and economic reform programs over the past four decades with mixed results, the argument for adopting private-sector practices has never been stronger. In reality, the private sector is flourishing while the public sector is in decline. This stark contrast has placed the government in a precarious position where change is no longer a consideration but a mandate.

Spawned by the Budget Control Act of 2011 and the American Taxpayers Relief Act of 2012, sequestration is the landmark measure of automatic, across-the-board spending cuts on government discretionary budget authority, anything

Congress funds each year through the appropriations process. Implemented on March 1, 2013, sequestration is expected to reduce government spending by $1.2 trillion between 2013 and 2021 (with the exception of "mandatory" programs such as Social Security and Medicaid). Many in the private sector are simply unaware of the impact these measures will have on the economy. Undoubtedly, sequestration will slow the US labor market recovery (at a cost of approximately 700,000 new jobs) and restrain the economic recovery by reducing the gross domestic product (GDP) by an estimated some 0.5 percent in 2013 alone (from 2.0 to 1.5 percent in 2013).

With sequestration's evolving rules of engagement, government leaders and decision makers will need to move beyond their traditional boundaries of influence to assess and consider how decisions they make will better serve the collective government ecosystem versus their own organization. This has not always been the case as many federal agencies have traditionally operated with a "silo" mentality, where organizational priorities are primary and everything else is secondary. The effort to scale back overinflated government spending plagued by years of bureaucratic inefficiencies, political fragmentation, and a behemoth Napoleonic organizational structure will surely lead to hard decisions in prioritizing the nation's requirements. The impact of sequestration is not only a public-sector challenge but also a private-sector challenge.

With few exceptions, defense and private-sector companies supporting the needs of federal organizations will also face an increasingly competitive environment in which the demand signal for their services has changed. In some cases, the demand signal and need for some services have simply gone away. For companies that fall in the latter category where the demand for their services has decreased, the window of opportunity for a well-planned transformation of business strategies and plans has passed, and they face some tough decisions moving forward. For those that remain, the demand for the "same or better services at significantly reduced rates" will only increase.

Individual agency priorities will inherently shift under the continuous pressure of competing priorities, reduced budgets, and increased scrutiny from government oversight committees.

Much like the impact game-changing innovations can have on competitors and service providers in a given market sector, philosophical changes in consumer priorities can have similar affects that challenge a leader's ability to posture their organization for mission success. As government officials and agency leaders continue to wade into a deeper state of introspection and reflection, new paradigms will form. Central drivers will be how these leaders posture their organizations to accomplish their assigned responsibilities, prioritize known or existing capability gaps in core competencies, and operate with a reduced workforce. With improved clarity of purpose and the acceptance of fiscal limitations, intrepid and visionary leaders will embrace this challenge with a goal of seeking new and innovative ways to create value, reduce risk, and build competitive advantage. In sum, government planners, programmers, and budget gurus across the federal enterprise will need to roll up their sleeves and sharpen their pencils in a concerted effort to redefine strategy, vision, and mission statements conducive to a resource-disparate future.

JCIDS and the Requirements Generation Process

Most program managers (PMs) are right from the operational environment (e.g., cockpit or support function) to leverage their knowledge and advice on addressing known and unknown capability gaps in the field. They are often teamed with experienced acquisition professionals (also referred to as PMs), who understand how the DAS, FYDP, JCIDS, and PPBE systems and processes interact to deliver capabilities-based solutions. This teaming or blended relationship leverages the operator's voice and experienced subject matter experts (SMEs) from the professional acquisition community.

Generation of requirements or needs occurs at any level within or outside an organization. From the field level, requirements often come via a service-specific requirements submission (document). This document is then vetted through the agency leadership levels until it reaches the organization's strategic planning, programming, and budget function for consideration and prioritization with other existing or known capabilities-based requirements. The individual services have some control over requirements and resource commitments within their assigned budgets. Those outside their respective budget or core competencies generally submit to the joint staff and JCB for review and further prioritization by combatant commands that link their respective core competencies across the services. Many questions often arise during the process that require continuous clarification and work at the major command and joint staff level. It is a tedious vetting process, often requiring thousands of dedicated government employees and defense industry contractors working in tandem with service providers around the clock to build a strong case (justification) for validating needs. There are several steps in this process as agency-specific requirements combine with other agencies' submissions that all travel in one submission to the POTUS for approval.

In the case of the military budget submission (requirements for new and existing programs are included), service-related requirements move through each service's secretary. Once approved by the service secretary, these budgets go to the secretary of defense (SoD). The SoD and his or her staff make a full review, prioritize requirements and programs, and subsequently submit the DoD's proposed budget to Congress and the Senate for review and approval. From this point forward, the president is responsible for the submission, with combatant commands and services standing ready to answer questions or provide further information to support their submissions. Again, you may feel you have a requirement or need, but until it becomes a program (e.g., allocated funding), it is not considered a program.

Based on each service's mentoring and succession plan (further discussion is provided in Chapter 8 on innovation strategies), future leaders and decision makers require a tour at a subordinate level of command, major command, service, or joint staff. This experience helps to introduce, train, and familiarize them with the planning, programming, and budget process. Unfortunately, not all excel at this type of work. Successful tours of duty at these increasing levels of leadership become milestones or demarcation points in their career progression versus a lesson in good product stewardship, judgment, and fiscal discipline. Each service has its share of trendsetting leaders who have vision, experience, and an eye for innovation. They have experience in transforming an archaic thought or new idea and transitioning that thought or idea into an innovative action or approach that transcends the boundaries of the known and accepted. Let us remind you that not every good idea or innovation is adopted or had a chance to succeed because of the lack of buy-in from the very core of professionals who were charged with implementing the process. Our position remains as follows: Unless you gain support and buy-in from those implementing any phase of a process or idea, you face an uphill battle.

Characteristic of autocratic societies that support democratic principles, decision making on proposed recommendations of requirements can be lost or reshaped as they move up the hierarchy in an agency or organization. There are many competing interests, and one might find they fall clearly within the buckets or categories established under an agency or organization's core competencies. However, depending on what level your beloved requirement (need) enters the vetting process, be prepared for heavy scrutiny and many questions on why your need is more important than others in the lineup. In truth, you may build what you consider is an airtight justification for why your requirement is more important than others are, but there is always a bigger picture. If you build the best case based on factual data, your chances of success improve.

Measuring defined requirements against long-term strategic planning and programming documents is generally the norm. This plan is a strategic document that attempts to take a 360-degree view of all aspects of an organization's key core competencies and the changes needed to posture the organization to meet current and perceived challenges in the future years. It is safe to assume each government organization has one. It sets the tone for categorizing and prioritizing most known requirements with the exception of unknowns that can surface from conflict, changes in core competencies, or global circumstances—all things that can drive a change or identify an unknown capabilities gap. These unplanned events can reshuffle global, national, and individual service priorities overnight. One needs only to consider the impact and fallout of sequestration to feel how priorities can change rapidly.

While assigned to the US Air Force's Air Mobility Command (AMC) headquarters at Scott Air Force Base, Illinois, it was a distinct honor working with AMC's finest as the branch chief for the Operational Support Airlift and Executive Airlift (OSA/EA) Requirements Branch. Anastacio "AL" Lambaria, a highly decorated retired officer and requirements guru, has an established reputation and enviable pedigree in mentoring countless young and seasoned officers, civilians, and contractors. One of his most prolific statements that brought everything into perspective in defining requirements often left many action officers speechless: "Requirements are needs, not wants." Considering one of the primary roles of an action officer at AMC, these individuals represent their aircraft platform community and support program requirements for prioritization and funding. No matter how important you (the user or action officer) might think a "requirement" is or might be, there is always the big picture your justification must consider. It reality, it must stand up to heavy scrutiny beyond the boundaries of your own influence and emotional attachment to the requirement.

Every organization has what is often referred to as the "1–n" list. The top priority (number 1) is the highest, and other requirements or funded/unfunded programs fall somewhere on that list. The charge to make a final determination of what a particular organization will allocate its funding to each year is a major event. There are many long hours involved, and committed professionals must be available on a moment's notice to support their assigned programs and requirements. Based on their own merit, programs and requirements drift up and down the 1-n list of the Air Force's requirements. In sum, this is an intuitive process and not a list set in stone or shared openly. We have all seen monumental programs funded for years cancelled (e.g., the presidential rotary or helicopter program cancelled by Secretary Gates after investing $2.3 billion in taxpayer dollars).

As this is a high-level personal opinion of how the process currently works, this was not intended to explain the entire requirements or acquisition process. There are many acquisition professionals who take their profession seriously, and they are willing to spend countless hours away from their loved ones to do the right thing; one of those professionals is US Air Force Captain Jeremy (Jerry) Baker, (PM, Air Force Material Command). A close friend and well respected in his own right, Jerry is a millennial by generation and has an incredibly inquisitive mind. He is known for not being shy about asking questions—a lot of questions—with the focus on understanding why things are done the way they are. Unfortunately, beyond Jerry's circle of influence is the ever-present bureaucracy that shuts down the questions, often leading to alienating the very voice that is in daily contact with the operator and end user of a product, service, or process. Many issues can kill innovation, but in this case, cultural boundaries built around position and title were the culprit. The higher one ascends in the hierarchy, the more power they have to advocate for or shut down innovation. As mentioned, if you want something accomplished, be prepared to put the idea in someone else's hands for advocacy as if it is their idea.

Systemic Leadership Challenges within the Requirements Definition Process

Every great organization develops leadership based on a plan of succession. Instrumental to that plan of succession is the tempo, titles, positions, increasing levels of responsibility, and performance at increased levels of responsibility. This does not rule out talent acquisition from the outside by any means. Outside talent can often be the catalyst for new ideas and a fresh perspective. They can also provide timely solutions to short-term leadership gaps and drive long-term benefits across an organization or agency. For example, military officers considered for positions of leadership succession typically find themselves in an 18- to 24-month leadership position cycle at varying levels within an organization. In this short period, they can often seek to change what is in place with an eye on improvement driven by an innate need to separate their performance from the pack and leave their mark. The idea of growing better leaders means job changes are frequent to foster a broader view of the complexities and roles within a particular agency. The reality is that having a title or position of responsibility can become self-serving, which tends to alienate the very workforce relied on to make the changes they profess will make the organization better. A common understanding is often the most difficult thing to achieve when leadership finds itself directing activities without gaining buy-in from their team—another hurdle for implementing innovation.

Instrumental to continuity within any defense-related organization are the government civilian, the embedded defense contractor, and the service providers with whom the organization has contracts for support, services, or products. Regarding embedded contractors, there was a time when the government functioned in a guarded and closed innovation environment where open and transparent information sharing could get you in a lot of trouble or cost you your job. In today's environment, that approach is changing to a more open innovation environment that creates partnerships and shares information to some level. Unfortunately, the remnants of the "old guard" mentality remain influential, but there is promise this closed-society approach is slowly being eradicated by a new generation of government employees, those who see open and transparent communication as a tremendous opportunity for achieving timely and innovative results. That said, we are not evangelists for the open sharing of all strategic or operational-level strategies; there are limits on sharing intellectual property (IP) and business intelligence (BI), and judicious discipline of what is shared remains a top priority.

In reality, government civilians and trusted embedded contractors provide the necessary baseline knowledge and continuity for critical programs that have a long timeline or take years to move through the PPBE, DAS, and JCIDS kabuki dance. When newly assigned leaders enter the scene, those who survive employ a mutual level of respect and appreciation for all and are often the best listeners. Unfortunately, some of these leaders make continuing assumptions based on their prior service, experience somehow trumps the experience and knowledge base of others, and they begin to reshape their organizations in a vacuum of their own thoughts. There is an unspoken rule in the military that guides new leaders within an organization: In your first 60–90 days, employ the 3 Ls (look, listen, and learn); in the last 60–90 days in your position, you should not introduce or implement new changes to your organization without a clear idea of what you are implementing.

Summary

Many may profess that the current relationship and each facet of the DAS, JCIDS, and PPBE need change, but few among that choir have the overarching knowledge and understanding to map a solution to make that happen. By nature, the DAS,

JCIDS, and PPBE processes are driven by many factors that simply require a concerted agreement on how to resolve the complex issues that prevent them from responsiveness to identified needs. At best, those close to this relationship see it as a "work in progress," a continuous journey and evolution toward a better state. What is important here is the individual nature in how people view innovation (as new, improved, or changed) that is of elemental importance. The online Defense Acquisition University holds the compendium of details on the full DAS process (http://www.dau.mil/default.aspx), and those committed to finding a solution to what appears to draw similar comparisons to the Minoan's quest to untie the fabled Gordian knot will continue to look for new ways to accomplish their missions.

As with private industry and innovation, the federal community continues to forge and update its basis for defining requirements in relation to its core competencies and known capability gaps. We must reiterate that innovation is driven by defined and real needs, not by emotions and desires (wants). Hence, the importance of having the right leaders and supporting cast at the helm, in the cockpit, or on the ground who have a good command of the "big picture" is essential.

Unlike the prescription N^2OVATE™ offers, some might argue that some programs within the requirement (need) structure do not have the right mix of players or keep the right players involved in an innovation through the entire life cycle based on their own succession plans. We acknowledge this approach is not conducive to the growth and succession models of how government agencies train future leaders. However, there are other models that many federal agencies might consider to build leaders and continuity of purpose. A continuous merry-go-round of leadership changes often complicates continuity of purpose and can definitely hamstring the innovation process. That said, government civilians (often retired veterans) experienced in acquisition and program management are called on to bridge knowledge gaps and the continuous leadership changes. We agree that not all leaders are cut from the same cloth and often desire to leave their mark through some type of change. However, this elevates the risk profile in most cases and can lead to potential waste, a lack of continuity, and in some cases, the derailment of a program or innovation. Sequestration is a major concern because it spawned a civilian drawdown where seminal program experience has and will continue to be lost, weakening the requirements definition process. At first glance, these may not appear favorable, but the changes they will drive will appear innovational.

Finally, although this is not a call for the government or DoD to change how it builds its future leaders, this is a statement that guiding change and transformation in any environment is no easy task when you change leadership every few years. Structure, knowledge, and principled leadership must offer some congruent bond. From our experience, some leaders trend toward throwing the baby out with the bathwater versus building on their predecessor's efforts or listening to the supporting players (i.e., government civilians) responsible for keeping the program stable as the leadership development cycle continues to run its course. In essence, we encourage the government to explore new options for building future leaders.

Discussion Questions

1. In your own words, explain the difference between a requirement (need) and a want.
2. What is a defining characteristic of a core competency? Can you provide one example of a core competency in your organization or agency?
3. What is the relationship between the Defense Acquisition System (DAS), Joint Capabilities Integration and Development System (JCIDS), and the Planning, Programming, Budgeting, and Execution (PPBE) processes?
4. Name three of the five phases of the DAS.

Assignment

1. Considering the DAS, JCIDS, and PPBE relationship, briefly discuss one example of where the N²OVATE™ methodology and system could be employed to improve an outcome. Complete some additional research on the three different systems and how they are designed to work in chorus with each other.
2. Faced with your own opinions and understanding of how succession plans work in your agency or organization, provide a potential change plan from your own experience that would minimize the impact of the leadership change while honoring the intent and understanding of the current leadership-building philosophy.

Chapter 6

Organizational Assessment: Is the Organization Ready for Innovation?

Introduction

For innovation to be successful, each federal agency and organization needs to understand its strengths and weaknesses. Accomplishing this recognition requires an introspective examination of the agency's or organization's preparedness for innovation. This assessment permits not only an understanding of where the organization is at present but also what it will take to become more innovation proactive. "Jumping in" to innovation without first understanding certain elements of the organization can prove disastrous. Innovation is more than a process; it is a strategic function dedicated to sustained long-term improvement.

The first step in developing an innovation effort that will succeed is in understanding the organization. A typical and normal reaction is that we all understand the agency/

organization, but from our own set of experiences and knowledge. Every other supplier, employee, and user (customer) has personal perceptions that govern their understanding of the organization. Required is an overall assessment—one that seeks the perceptions of others to establish a baseline. The process begins at a high level initially, and then subsequent evaluations will focus on the organization's culture and work environment.

This chapter focuses on assessment and evaluation. Once the organization completes the assessments, it will better understand how to implement innovation for sustained success. We recommend simple assessment tools given that the information provided to federal agencies and organizations is a unique and descriptive examination of the organization from the perspective of innovation. The first assessment focuses on the organization's or agency's propensity and openness to innovation.

Organizational Evaluation

Before discussing initiation, consider the readiness of the organization for innovation. The simple assessment instrument will provide a quick overview of "organizational readiness for innovation." Complete the assessment and score each question. Select executives, directors, and managers within your organization or agency as a sample group. Try for a minimum of 20–30 individuals. Check the box that best applies. Realize that the scores represent a perception of overall readiness. Sum the scores to determine the overall readiness score.

Initial Assessment: Organizational Readiness for Innovation

Please answer each question truthfully, selecting the most appropriate response:

1. Is there a specific strategy for implementing and maintaining innovation?
    ☐ (3) Yes, fully functional ☐ (2) Yes, marginally functional ☐ (1) No, or I do not know
2. Is leadership open to changing how it develops a strategy for innovation?
    ☐ (3) Open ☐ (2) Somewhat open ☐ (1) Generally not open
3. Does leadership discuss the benefits of innovation freely with associates and users?
    ☐ (3) Always ☐ (2) Occasionally ☐ (1) Rarely or never
4. Do the mission and purpose statements support an environment for innovation?
    ☐ (3) Always ☐ (2) Occasionally ☐ (1) Generally not
5. How well does leadership manage the organization for innovation success?
    ☐ (3) Well managed ☐ (2) Sporadically good management ☐ (1) Frequent missteps
6. How would you rate the preparedness for conducting and sustaining innovation in your organization?
    ☐ (3) Fully prepared ☐ (2) Marginally prepared ☐ (1) Unprepared
7. Rate how "innovative" the organization is at present.
    ☐ (3) Organization is always in "catch-up" mode
    ☐ (2) Comparable to like agencies or organizations
    ☐ (1) Struggling to keep up with other agencies or organizations
8. What is the major reason for innovation in your organization? (Check all that apply.)
    ☐ (3) To provide value ☐ (2) To increase efficiencies
    ☐ (1) To reduce overall costs
9. How would the organization rate its ability to implement new ideas?
    ☐ (3) Ideas directly influence new products and services ☐ (2) Ideas occasionally influence new

products or services ☐ (1) Ideas rarely create new products or services
10. How resistant to change are the employees of the organization?
☐ (3) No resistance ☐ (2) Occasional resistance ☐ (1) Employees resist change
11. Is Innovation one of the organization's stated values or goals?
☐ (3) Yes, discussed regularly ☐ (2) Occasional references ☐ (1) Not a part of the organization's values and goals statements
12. How would innovation add value to your agency/organization?
☐ (3) Significantly add benefit ☐ (2) Minor benefit added to the organization ☐ (1) Do not know

Total organizational readiness score (add scores for questions 1 through 12): _____

Interpretation

A simple evaluation will not be 100% accurate. The intent of these questions is to inquire if the organization needs to develop an innovation management strategy. Few agencies or organizations have developed innovation as a strategy. Most agencies and organizations know they need innovation but are unsure of how to implement the concept. Overenthusiastic scores would mask the need for this strategy; low scores would suggest that the organization is in need of innovation. Therefore, use the following guidelines to assess your agency's or organization's openness and preparedness for innovation:

Scores of 31–36 = The organizations are well prepared.
Scores of 25–30 = Opportunities for innovation exist; improvement is needed.

Scores of 18–24 = Innovation is not a strong value; innovation is needed as a working strategy.

Scores less than 18 = Innovation is unimportant or the agency or organization needs a redirected focus. Lower scores indicate the potential for improvement and further delineation of program needs.

Actions

Scores should realistically be in the 18–25 range. Higher scores indicate that an agency or organization is producing innovations on a regular basis; further evaluation is not required. Lower scores signify the need for innovation and the need for change. A proactive approach is to examine the scores and decide whether

1. The organization is not ready for innovation.
2. The organization is ready for innovation.

For those not ready, the choice to proceed or retreat is evident. Proceeding forward requires an evaluation of those lower scores; we offer the following questions for consideration by those agencies and organizations that desire change:

1. Are these short-term or long-term concerns?
2. Are these structural (policies and procedures) or subordinate (leadership, climate, history, etc.)?
3. What are the resource requirements?
4. Is the organization ready to change?

Answering these questions will provide a realistic evaluation of what it will take to initiate a culture that supports innovation. Assume the culture can change as the innovations become more commonplace. Recognizing the need is crucial for developing an effective strategy and a successful management process.

For those ready for innovation, we suggest proceeding with the next round of assessments and evaluations (e.g., work environment, value assessment, etc.). The goals and objectives of these assessments and evaluations are a deeper understanding of the agency or organization's strengths and weaknesses and propensity for evolving their current culture and structure to one that is more innovation-centric. Becoming an innovative agency or organization is a journey. Central to that journey is a concerted effort to maintain open and responsive channels of communication up, down, and laterally within and outside the organization. The promise of success is never a guarantee, but the aperture of opportunity expands greatly when accomplishing assessments with objectivity, full disclosure, and in an open and honest manner. In summary, assessment instruments provide more focused information for decision makers to consider. Part of the process of implementing an innovation strategy is the alignment of an organization to a proactive status.

This simple assessment is only the initial evaluation of the agency's or organization's posture regarding innovation. In no way does it provide the basis for an action plan or strategic decision. The next step is to evaluate how executives, directors, managers, and employees judge their agency's or organization's innovation success.

Evaluating Innovation Success

It seems easy to judge success when comparing results from our own perspective. Yet, organizationally, success can come in many forms and provide a wide range of benefits. Understanding how success affects and what success means to employees and management is critical to establishing goals and objectives as you plan a strategy. Far too often, management narrowly frames innovation success into one or two outcomes (objectives), missing more critical or value-added organizational objectives. A key element in developing a successful strategy for innovation is planning to address multiple objectives beyond those normally associated with improvement (lower costs, improved effectiveness, and efficiencies). Additional objectives permit the definition of innovation to expand beyond an agency or organization's traditional approach, permitting public service–related organizations to experience innovation success with less-tangible measures. For example, improving communication effectiveness may have a more direct impact on the organization than a project that reduces costs.

Therefore, before embarking on an innovation effort, consider how associates define successful innovation outcomes. Be open and willing to consider a wide variety of responses confirming our assertion that innovation begins with the individual. Distribute these few questions to a variety of employees from the executive level to where the rubber meets the road—at the implementer or worker level.

1. What are the critical elements needed to ensure innovation success? (Check all that apply.)
    - ☐ Proactive leadership
    - ☐ Employee collaboration
    - ☐ Effective communications
    - ☐ An effective method or process to execute innovation projects
    - ☐ Available funding
    - ☐ Effective technical support from (research and development, engineering, information technology, human resources)
    - ☐ Clear objectives and outcomes
    - ☐ Effective policies and procedures
    - ☐ Understanding of what the internal and external customers (users) need

☐ Satisfied internal and external customers (users)
☐ Empirical data
☐ Technology, product, or service performance
☐ Other; please specify ____.
2. How will innovation benefit your division/department? (Check all that apply.)
   ☐ Improved communications
   ☐ Improved collaboration within the organization
   ☐ Improved operations/services
   ☐ Improved user (internal and external) satisfaction
   ☐ Greater efficiencies
   ☐ Reduced costs
   ☐ Less waste and errors
   ☐ Increased productivity
   ☐ More effective workforce
3. Judge the value of the innovation opportunity on (check all that apply)
   ☐ How new or novel it is
   ☐ How many people it involves
   ☐ How much it costs
   ☐ How long it takes
   ☐ How much improvement occurs
   ☐ How much benefit it provides
   ☐ The extent of the change that occurs

Interpretation

A variety of responses confirms our assertion that innovation means many things to many people. Assessing the organization will assist in establishing long- and short-term objectives at the strategic, operational, and tactical levels as well as broaden the scope of innovation. Use the results from this assessment to help define the expected outcomes of an organization-wide effort.

Often, this information leads to a filtering process intended to logically group (by function or topic) objectives that reduce their number. Be careful to include measures that relate to people, policies, procedures, processes, products, or services and select objectives that are easily measured. For instance, it would be great to have an objective that states that employee morale will increase with innovation. The problem lies in measuring morale objectively and accurately, which is situation dependent. The goal here is to select measures that define the innovation outcome.

Actions

This simple survey does not require any calculations. Collect the data and graph the responses in Excel for each of the three questions. Expect your responses to be as varied as shown in Figure 6.1. Conducting an 80/20 rule analysis would consider only those questions with a score of 9 or higher (these are considered possible success criteria) (Figure 6.1). The important element is the diversity of responses that reinforce the concept that measuring innovation success covers a wide range of criteria from organizational to human issues.

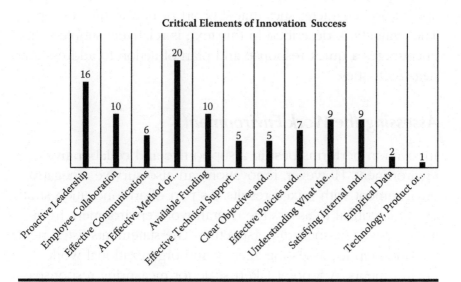

Figure 6.1 Critical elements of innovation success: question 1.

At this point in the assessment phase, consider such distinctively human issues as cultural ones (associated with the agency or organization) and an evaluation of the work environment. The focus is on identifying traits of the agency or organization that support or discourage innovation by examining the culture and its influence on leadership, management, and employees. This does not negate the cultural and ethnic diversity so necessary for innovation. Rather, it assesses the prevailing organizational culture and its ultimate influence on the success of any innovation effort.

Evaluating Agency and Organizational Culture

Federal agencies and organizations have developed a unique culture based on their mission and purpose. This culture influences all aspects of management, from decision making to operational efficiencies. Often, the culture is opaque to other agencies or organizations that must interact or depend on the actions and responsibilities of the host organization. Therefore, it is important when beginning an innovation effort to evaluate cultural characteristics that support or negate the concept. The evaluation, described in this text, is relatively simple so it encourages a quick response and plan of action to address any negative issues.

Assessing the Work Environment

Innovation performs best in a supportive and collaborative environment. However, innovation can also survive a negative environment with mixed results. As part of an innovation strategy, it is critical to assess the overall work environment from an employee (associate) perspective. The statements listed next provide help for assessing agency and organizational work environments. A 5-point Likert scale for measuring responses follows. Rate these statements objectively and honestly by selecting the level of agreement that best fits your feelings at the time of this assessment. 1, strongly disagree; 2, disagree; 3, neither disagree nor agree; 4, agree; 5, strongly agree.

- ☐ The work environment supports creativity.
- ☐ I have confidence in my own abilities to solve problems.
- ☐ My workplace provides me with challenges.
- ☐ The workplace enables me to be creative.
- ☐ The work environment is open to new ideas.
- ☐ There is a sense of cooperation among employees.
- ☐ Management rewards improvements.
- ☐ Work demands are not overburdening.
- ☐ My workplace encourages change.
- ☐ Mistakes are not sources of anxiety or ridicule.
- ☐ Trust is a value of my workplace.
- ☐ Collaboration is encouraged when problems arise.

Depending on the size of the agency and organization, distribute this assessment to a large number of employees (associates) for completion. For organizations under 50 employees, select 10; between 50 and 100, select 25; between 100 and 500, select 100; and so on. Stress the need for objectivity and honesty; these are best accomplished when this assessment is anonymous. After collecting the completed assessments, calculate an average score per question. Coupled with a demographics section, comparing information such as employee (associate) tenure with the organization, job function, gender, and age (or generation) can assist in better identifying where your innovators are within the agency and organization. Phrasing the instrument in a positive manner eliminates confusion. The goal is not to compare one individual versus the next; look for patterns in groups or clusters of individuals. Finally, calculate an overall score (all 12 statements) per respondent. This will provide information on the respondent's overall assessment of his or her work environment.

Interpretation

Consider this assessment of the work environment as a measure of employee satisfaction. Avoid the tendency to look for and highlight large differences. Subtle indicators will provide information that is far more useful. Previous research has demonstrated that groups often rate the 12 statements as a single concept, suggesting all issues hold the same importance. Overall scores should range between 4.00 and 4.50. Focus on the variation between statements: Large differences in how a respondent scores a work environment statement indicates a varied work environment (both positive and negative). The information is perceptual, and this could easily change when encountering a negative situation.

Actions

Associates (employees) can easily make or break an innovation effort. If this group shares a positive view, then this group is ready for innovation implementation. A neutral score highlights the need for action and a negative score the need for patience and introspection as the organization may not be ready for innovation. After the organizational and success criteria assessment is completed, the next step is to evaluate the operational processes for innovation readiness. Using a combination of instruments and philosophies, such as those of Higgins (1995) and Brand (2010), the organization can assess its state of readiness for its innovation journey—how "ready" the organization is for an innovation management process.

Innovation Readiness

If the organization is open to innovation and can readily define a broader definition of innovation success, then further assessment can continue. The next assessment we offer is the innovation readiness (IR) assessment that evaluates the seven critical components of a successful Innovation Management (IM) program (explained in further detail beginning in Chapter 8). These components establish the essential steps required in evaluating and implementing innovation projects. The associated 5-point Likert response scale for each component or question is as follows: 1, strongly disagree; 2, disagree; 3, neither disagree nor agree; 4, agree; 5, strongly agree.

Step 1: Needs (Requirements) and New Ideas

Please select the best response that matches your beliefs about innovation needs and new ideas:

1. Managers are open to new concepts.
2. Employees are naturally creative.
3. What the user needs is important.
4. Unhappy users require immediate attention.
5. Employees can easily bring new ideas to their supervisor.
6. Needs must add value.
7. Users generally need what is often not available.
8. The organization rewards new ideas.
9. The agency/organization/division/department is open to new ideas.
10. The organization has strong ties to its internal and external users.

Step 2: Nominate and Normalize

Please select a response that best describes your beliefs regarding the following statements:

1. The organization chooses the best people to work on improvement teams.
2. Job responsibilities are clear to employees.
3. Team leaders or champions lead project teams.

4. New projects generally do not disrupt operations.
5. Things do not run better when left alone.
6. The best people work on new projects.
7. Keeping a low profile rarely works for employees.
8. Improvement teams require the best employees.
9. The culture encourages employees to innovate.
10. Employees want to help the company succeed.

Step 3: Objectify and Operationalize

Please select a response that best describes your beliefs about objectives and operations:

1. Employees generally support agency/organization/division/department objectives.
2. Innovation rarely interferes with operations.
3. Objectives are clearly communicated with all employees.
4. Projects need a specific objective or goal.
5. Understanding process and product requirements is critical.
6. New technology drives innovation.
7. Improving something step by step is critical.
8. The organization needs innovation training (and certification).
9. Limitations never reduce our success.
10. When change is positive, it can be innovative.

Step 4: Verify and Validate

Please select a response that best describes your beliefs regarding how innovation is measured and evaluated:

1. The organization readily communicates innovation success.
2. When a project fails, the reasons for its failure are clear.
3. It is important to measure performance.
4. Innovation efforts produce best practices.

5. Quality checks always occur before releasing a product or service.
6. Innovation provides real value.
7. High quality standards are important.
8. Customers and shareholders drive innovation efforts.
9. New technology drives innovation.
10. When the situation changes, the organization can react.

Step 5: Align and Adapt

Please select a response that best describes your beliefs regarding how people align and adapt to innovation:

1. A champion or team leader manages innovation projects.
2. Management is accountable for all innovation success
3. There is a process for creating new products or services.
4. There is a high level of trust.
5. This organization is a good place to work.
6. Management looks for ways to improve products and services.
7. Employees adapt quickly to change.
8. Teams work best when working to the same objective.
9. Decisions regarding innovation are never made hastily.
10. The work environment is informal.

Step 6: Track and Transfer Performance

Please select the best response that matches your beliefs regarding how projects are tracked and transferred:

1. The organization dedicates additional resources for innovation.
2. Spending on innovation projects is tracked and monitored.
3. Performance is continually monitored.
4. Employees receive a reward for innovative ideas.
5. There is always a risk that the innovation will not work.

6. If a project fails, no one is blamed.
7. Transfer time is short between innovation and operations.
8. New ideas are easily accepted.
9. Resources are present for innovation today.

Step 7: Evaluate and Execute

Please select the best response that matches your beliefs regarding a project decision:

1. There is follow-through on new projects.
2. External and internal users judge our products and services for value.
3. Checks and balances ensure high quality.
4. The organization shares best practices with employees.
5. Employees easily adapt to new changes.
6. Managers ask for feedback.
7. Data is an integral part of most decisions.
8. The organization documents project progress.
9. Innovation is part of the organization's strategy.
10. The organization is ready to become an innovation leader.

Scoring

The IR scoring portion of this instrument represents a method of assessing the organization's strengths and weaknesses. Each statement is a characteristic of innovation (seven major components). When a respondent evaluates a characteristic favorably, identify it as a "strength." Conversely, when a respondent evaluates a characteristic negatively, identify it as a "weakness." The IR position is the sum of positive scores. To score the IM portion, convert all statement scores into one of three IR score categories:

1: When the respondent scores a 4 or 5 (a positive response)
0: When the respondent scores a 3 (a neutral response)

–1: When the respondent scores a 1 or 2 (a negative response)

Consider Table 6.1 as an example.

Interpretation

There are seven total IR scores, one for each component. The sum total is the IM score. For those agencies or organizations in advanced stages or cultures of innovative performance, it is prudent to recognize that achieving a score of 70 would be difficult, as it would for those on the opposite end of the innovative performance scale to score –70. Typical scores will range between 20 and 50. A common mistake is averaging the scores. Do not average scores as the overall average could be close to zero and negate the purpose of the assessment. Sum the scores for each N²OVATE™ category. Examine the range or distribution of scores. The higher the score is, the better prepared the organization is for innovation at the operational level. Lower scores indicate targeted areas that require improvement or modification before implementing an innovation strategy. Large deviations between scores indicate a lack of consensus regarding preparedness (Figure 6.2). Total scores for the 50 respondents demonstrate this lack of consensus. Rather than this being considered as a negative indicator, we recommend additional steps be taken to determine the reasons for the disagreements.

The IR instrument can also assess a present process or innovation management system. Use the scores more as indicators than solid, empirical numbers.

To determine which N²OVATE™ category is most (least) capable, average the scores across the number of respondents (the sample). Construct a "radar chart" (Figure 6.3) across the 7 categories (Seven Steps). Remember that averages will be close to zero. A positive response indicates compliance with the process step (discussed further in Chapter 9). A negative

Table 6.1 Example of IR Instrument Scoring

Needs and New Ideas Component	Strongly Disagree 5	Disagree 4	Neither Disagree nor Agree 3	Agree 2	Strongly Agree 1	IM Score
1. Managers are open to new concepts.			X			0
2. Employees are naturally creative.				X		1
3. What the user needs is important.				X		1
4. Unhappy users require immediate attention.					X	1
5. Employees can easily bring new ideas to their supervisor.			X			0
6. Needs must add value.			X			0
7. Users generally need what is often not available.				X		1
8. The organization rewards new ideas.				X		1
9. The agency/organization/division/department is open to new ideas.		X				−1
10. The organization has strong ties to its internal and external users.			X			0

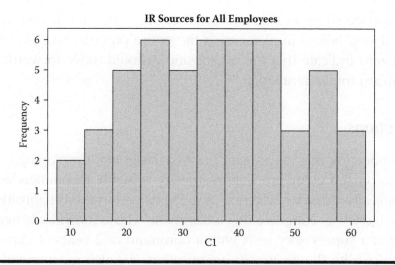

Figure 6.2 Total IR scores for 50 respondents.

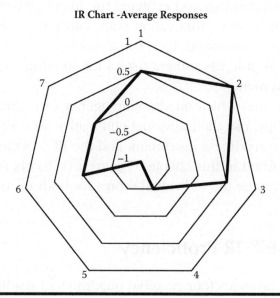

Figure 6.3 Radar chart of respondent averages.

response suggests that additional effort is required to improve that step before implementing the entire process. Scores close to zero indicate that the organization should move forward toward implementation.

Actions

Conducting this assessment prior to designing, developing, and pursuing any innovation strategy is highly recommended. Carefully consider selecting individuals within the organization or agency to complete this assessment. These individuals must be at a supervisory level with a minimum of 2 years of experience with the agency or organization. We also recommend that the organization provide training with this assessment before it is distributed and the resulting data analyzed. We must emphasize that the use of this assessment is extremely beneficial as a precursor to implementing an innovation strategy. The assessment provides valuable insight and information on the perceptions of those individuals responsible for implementing, maintaining, and evolving the intent of your organization's or agency's innovation plan. Further, consider the value in the scores rather than the absolutes. The IR instrument is a versatile assessment useful for auditing and continuous improvement.

To determine which quadrant an agency or organization would occupy, use the individual cumulative scores from the organization readiness instrument and the IR proficiency instruments described in this chapter (Figure 6.4). To obtain an overall score, average the individual scores for both instruments.

N²OVATE™ IR Proficiency

Figure 6.4, provides four possible quadrants of readiness. Quadrant 1 (Q1) delineates a project management focus; quadrant 2 (Q2), the optimum situation; quadrant 3 (Q3), the

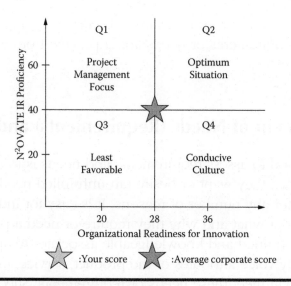

Figure 6.4 Innovation organization score.

least-favorable condition; and quadrant 4 (Q4), the conducive culture state or condition. From our perspective, Q3 is the least-favorable situation. Q1 has the correct focus but not all the ingredients for success, whereas Q4 is culturally adept for innovation but will need to supplement this with a project management approach. Ideally, Q4 represents the best situation because it exhibits a conducive culture with a process-focused mentality. Q4 contains all the elements (a conducive culture and project management focus) for innovation to succeed. The intent of this "grid approach" is for agency and organization leadership to use the grid as a visible indicator for improvement initiatives rather than course correction.

Innovation Processes and Solutions, LLC (IPS) customizes all assessments to a specific user or customer need. Assessment at this level questions the overall viability of a project to achieve its objective. This assessment begins at the first stage with the evaluation of requirements (also synonymous with the term *needs* throughout this book). Our assessment is that requirements or needs drive innovation. Unfulfilled needs or requirements can represent a longing or desire for something not yet available or cost effective.

Therefore, success with innovation requires an understanding of these requirements or needs and a process from which to fulfill them.

Assessment of Needs (Requirements) and Values

One method to assess requirements and needs is to ask individuals what they want or desire (an unfulfilled need) but cannot have for any number of reasons. It is easy for individuals to talk about what and why they require or need a particular item. Determined and knowledgeable associates (employees) will readily voice attributes of the product, service, or technology that will meet these needs or requirements. Successful leaders and decision makers are often good listeners, and they leverage these attributes as essential parameters for validating requirements and needs.

Identifying those individual attributes that directly affect the judgment of innovation is paramount. For example, a typical attribute is an evaluation of how an item satisfies the individual. Ultimately, satisfaction is the quintessential difference between perceptions and expectations. If perceptions (real-time experiences) exceed expectations, that is positive satisfaction. When perceptions are less than expectations, then we feel it is classified as no satisfaction. Further, satisfaction is an excellent barometer for a momentary evaluation of user or customer sentiment. That said, satisfaction can be fleeting and change radically with a negative experience.

Based on our views of satisfaction as a viable measurement, we propose that a better measure of overall sentiment is attitude. One such attribute is the attitude the customer (user) develops with continued use of a product (e.g., technology) or service. We also feel that attitudes are more concrete than perceptions of satisfaction. Individual attitudes develop from our own personal experiences and ongoing beliefs. We are not the first innovation professionals to advocate for or cite the need for measuring attitudes across an organization as they relate to innovation. Attitudes directly relate to the value people place on an outcome, product, or service. Thus, customers (users) tend to perceive and evaluate the value of an outcome, product, or service received against the price they have paid for that outcome, product, or service. Subjectively, the term *price* not only refers to cost and revenue but also can relate to time allotment, intrinsic or extrinsic value, energy expended, time invested, and so on.

For the federal sector, measuring value as a key attribute of innovation has a prominent place in defining the meaning of why something is important. Value has a longer-term perspective; it can easily drive behaviors related to buy-in, purchasing, acquiring, storing, protecting, investing, and so on. As the definition of value can differ from one person to another, we use value to mean something that the person or organization identifies as important. In the federal sector, defined needs (requirements) drive value, address identified capability gaps, and drive acquisition decisions and innovation opportunities; collecting and refining this information is critical for sustained innovation success.

For the federal sector, value can take on multiple meanings. Knowing what individuals value provides doorways to help agencies and organizations discover and define the unfulfilled needs and requirements. A value that remains unmet (not realized) influences needs and requirements. Figure 6.5 identifies a simple process of collecting, filtering, and organizing this information. A simple round of questioning provides sufficient support for developing the need into an innovation project. Understanding what the stakeholder values is integral to understanding what constitutes success. This information becomes the cornerstone of critical business intelligence (BI) and a measurement for future comparisons. The information also solidifies a basic innovation requirement: needs and values. The values will provide measurements of success, a guidepost to determine the degree and amount of

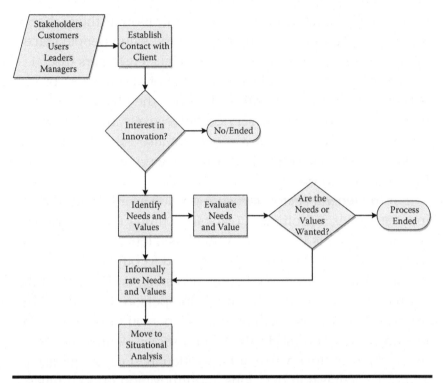

Figure 6.5 Assessing needs and values.

improvement. The next chapter discusses situational analysis as part of collecting BI used for diagnostic purposes.

Summary

Assessment is key in understanding the present state of an organization. Assuming an organization is ready for innovation, management is both bold and potentially reckless. The assessment instruments provide assistance in understanding the present state of affairs in the organization as it specifically relates to innovation. Assessing the present state of innovation readiness provides information on potential "holes" or flaws within the existing strategy. These "gaps" provide a potential for improvement or an indicator of possible trouble. Of course, these are subjective by nature but provide an information consensus. We use these to design a management strategy that highlights the strengths and weaknesses of an organization-wide innovation approach.

Interpret the information for the good of the effort. Use this information to build or refine an innovation strategy (see Chapter 8) and for planning purposes. Examine and interpret the data with the objective of improvement, stabilization, and sustained success. Leadership plays a key role in developing and supporting this philosophy.

Discussion Questions

1. Why is the concept of assessment critical for sustained innovation success?
2. Why assess an organization prior to instituting an innovation strategy? What barriers could prevent an honest evaluation?
3. Describe three learning points from an organization that is not "innovation ready."
4. How would you advise an organization that was not yet ready for the N²OVATE™ process?
5. How critical is it for an organization to know its needs and values before embarking on innovation?

Assignments

1. Explain how you would present information to senior management regarding an organization's innovation readiness score. If the scores were low, what would be the next steps the organization should address?
2. Describe a simple process to create a communication network in your organization that promoted honest exchanges, valuable feedback, and effective two-way communications.

3. Complete a needs-and-value assessment for your organization. Identify the synergies and differences. Choose one difference and decide on a remediation strategy to alleviate or mitigate the problem.

Chapter 7

Organizational Diagnostics: Through the Looking Glass

Introduction

Before implementing an innovation strategy, the organization needs to diagnose its present situation. Like Alice in Wonderland, the organization must peer through the looking glass. Often, the need for innovation arises from an existing program or possible opportunity. Conducting the review occurs after the assessment phase is completed but before identifying the best innovation strategy for implementation. Figure 7.1 details the steps involved. The technique begins when an organization (agency) expresses interest in working with Innovation Processes and Solutions, LLC (IPS). Prior to performing any work, we first analyze the present environment, including an assessment interpretation. Analytics provides the information collected; diagnostics provides the interpretation and follow-through.

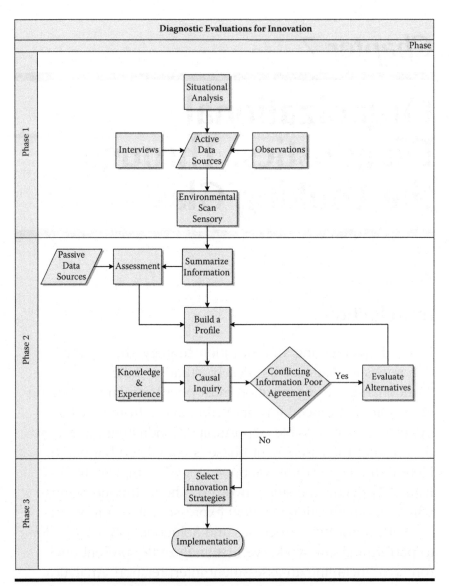

Figure 7.1 Diagnostic evaluations for innovation.

Why Perform Diagnostics?

Consider the situation when you have an illness that requires the attention of a physician:

> You make an appointment to meet with the doctor to discuss your symptoms and find a remedy. When you enter the office, the doctor or nurse begins by collecting your vital signs, interviewing you as the patient, and recording information for the physician. We call this *situational analysis*. On entering the examination room, the physician performs a visual survey of your condition (an environmental scan), asks questions (begins a causal inquiry), and examines your chart (examines passive data), performing further examination (collecting active data). Using personal knowledge and experience and the prevailing information (seasonal, common symptoms, etc.), the physician makes a diagnosis. If conditions warrant (e.g., there is conflicting information), the physician may do a "what if" analysis using a process of elimination to reach a diagnosis.

We use a similar process to understand why innovation can be beneficial to an organization and use this information to suggest a "best" strategy. Unlike those who only sell a "cure" (a method), at IPS we determine the best method (strategy) based on the organization's present situation (needs). This may mean creating a one-of-a-kind (proprietary) strategy for each organization.

To complete the evaluation accurately, use an evaluator with experience working with or at multiple locations. The best evaluator is an intuitive person, one who would rather understand the existing organization rather than judge it for compliance. The evaluator has to demonstrate a professional attitude, demeanor, and respect for the organization.

Begin the process by completing a situational analysis. Follow the process (Figure 7.1) until there is enough information to choose an appropriate innovation strategy.

Diagnostic Elements

Situational Analysis

Performing a situational analysis consists of examining the elemental units that describe the situation of interest. Identify and describe the following:

- Location (where the situation takes place)
- The "event" (underlying purpose or reason for the event)
- Outcome (presumed and actual)
- Context (in which the situation exists)
- Assumptions (made and presumed)

When examining an organization for its present and future innovation capabilities, consider not only the encounter (event) but also the preconceived notions the evaluator has before assessing the situation. This includes the assumption made before beginning the evaluation. Evaluate with the following context in mind (the pressure on the client to demonstrate positive results), where the evaluation takes place (location), and the expected/actual outcome. Often, evaluators orchestrate their reviews based on what they expect and want to occur. This "bias" will cloud the judgment of the evaluator. Complete a situational analysis without preconceived notions. The experience of the evaluator is critical.

Consider, as an example, the situational analysis performed by a police officer when evaluating a minor traffic accident. Often, there are two or more conflicting accounts of the accident. The officer must evaluate the situation without bias to make a report. The officer must take into consideration the location, existing circumstances, driving conditions, time of day, and more. The situational analysis determines the prevailing conditions in which the accident occurred. Next, evaluate the "environment" in which the organization operates.

Environmental Scan

An environmental scan involves examining the organization from a cognitive and sensory approach (Aslakson et al., 2014). The evaluator examines the surroundings to develop a profile of the environment in which the innovation does or could occur.

The environmental scan consists of a

- Visual assessment (place, presence, posture, people [e.g., body language/facial expressions])
- Verbal assessment (tone, clarity, communication effectiveness [talking vs. listening])
- Cognitive assessment (knowledge, experience, preparation)
- Setting assessment (atmosphere, mood)
- Emotional assessment ("feel," comfort factor, emotional state)

A visual assessment is critical to establish a baseline. Place is the physical location (i.e., office, restaurant, pub); presence is the authority, rank, or title of the people attending; posture involves physical items (seating arrangements, furniture); and the people dimension involves body language, facial expressions, eye contact, and the like.

Next, evaluate the verbal cues received through the encounter. The content should be clear, direct, and easy to understand. Tone measures rhythm, style, and intent (friendly vs. confrontational). Finally, evaluate the effectiveness of communication (value, delivery, content).

Every environmental scan assumes that we have some knowledge and experience with the event. Knowledgeable and experienced evaluators are prone to ask more questions. Those with less knowledge (experience) will listen and learn. A good rule of thumb is to include both approaches in an interview.

The setting is the characteristic that evaluates the atmosphere in which the event will occur, including items such as setting (which could be positive, neutral, or negative) and overall "mood." If one of the members of the encounter is uncomfortable, this will affect the result and could sabotage the meeting. If this is detected, try to diffuse with appropriate humor, a welcoming demeanor (invites a person into the encounter), or identification and elimination of the reason for this discomfort.

Finally, the emotional component is the evaluator's own "state of mind" in entering the event. Concerns such as preparedness, anxiety, and behavior (extrovert vs. introvert), trust, intuitiveness, and confidence are critical. Because no one is perfect, prepare a strategy to deal with emotional issues, both for the evaluator and for the other persons involved in the encounter. Next, consider the type of information (data) collected.

Active Data (Information) Sources

Both situational analysis and environmental scanning result in a large amount of data (information). Active data is real-time information, collected by the evaluator in real time. Active data consists of counts and categorizations. An example of active data is the information collected from an environmental scan. It can include observations, experiences, perceptions (emotions, opinions, beliefs), and measurements. The goal is in finding a pattern and linking the data to a particular attribute. The information is more descriptive than strictly numerical (quantitative) information. For example, temperature can be both quantitative (73°F) and qualitative (pleasant or mild). This type of data reduces inconsistency but is open to the effects of bias. This information source provides an excellent analysis of short-term events.

Active data contains valuable information. Websites, such as Facebook, permit and encourage expressions, observations, and a forum for experiences. It is generally unstructured data, but data scientists are actively tracking this data for economic and competitive information. For innovation evaluations, capture and record information directly related to the situational elements and environmental scans.

Passive Data (Information) Sources

Passive data is time-dependent information not collected in real time by the evaluator; analysis of the data can occur in various formats. Assessment-type data is passive in that the respondent (data source) is not directly in contact with the evaluator. Passive data reduces bias because the evaluator cannot directly influence the source. This information source provides longer-term information. Data sources can consist of individual respondents, databases, organizational records, government statistics, and so on. Passive data can be descriptive, measured, and categorized. It is the most frequently used; however, we use active data (short term) for most decisions.

The usefulness of passive data decreases with time. It can provide nearly useless information if the intention and purpose are misaligned. If measured incorrectly, it can cause problems or catastrophes. Relying on passive data only is dangerous, and it can distort the true reality.

Assessment instruments (surveys) collect passive data. This information helps to determine the "larger" picture, provides a history of the occurrence, and describes what individuals think and act on as a group.

Summarize the Information

Summarizing the observational and situational analysis data is key to understanding the organization in real time. When summarizing the data, consider the following:

- Patterns and trends (in behavior and actions)
- Verbal emphasis (what employees talk about)

- Convergent and divergent views (what individual agree on or disagree with)
- Attainable versus unattainable goals/objectives
- The importance (or irrelevance) to employees/management
- Similarities and differences (personality types, work ethic, loyalties, etc.)

Summarize the information into a usable format. Continue to update while continuing to work with the customer (user). One suggestion is to build a profile.

Assessment

Figure 7.2 displays the actual process associated with creating an assessment tool as discussed in Chapter 6. This assessment provides information when building the profile. To summarize and interpret the data, the evaluator incorporates both the situational and the environmental scan data. This identifies critical elements, such as

- What employees value.
- How employees best understand innovation.
- Whether the environment will support an innovative culture.
- How innovation benefits the organization, customer (user), and employees.

The next step is to design an appropriate data collection instrument and then finally to collect and analyze (synthesize) the data. This information becomes integral to building the organizational profile.

Building a Profile

Build a profile of the organization. Use the information collected from the situational analysis and environmental scans. An organizational profile is a short but concise description of

Organizational Diagnostics: Through the Looking Glass ■ 137

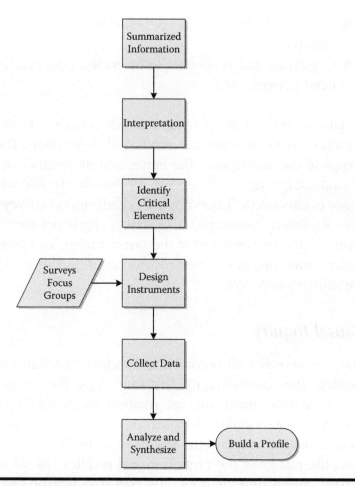

Figure 7.2 Assessment process.

- The organization's health (growth, stability, customer/user satisfaction, cost containment, resource usage, mission/vision alignment, etc.)
- The amount and success of value adding (what value is added and by whom)
- Leadership and decision-making success/failure
- Employee/management relations/communications (including aspects of trust, commitment, honesty, engagement, etc.)
- Influence of customers (users) and resource providers (i.e., stakeholders)

- Innovation efforts (value/benefits) (what worked, what failed)
- Organizational properties (efficiencies, effectiveness, products/services, etc.)

This is an example of what a profile contains. Each organization has its own unique needs and description. The more complete the description, the better and more efficient choice of strategy for sustained success. Update the profile when major events occur. Knowing the organization is a key to success. It prevents missteps (false starts); aligns personnel; demonstrates the competency of the organization; and promotes better communications. Security and confidentiality of the data are primary concerns.

Causal Inquiry

The evaluator is well prepared to conduct a preliminary causal inquiry after completing the first four steps. The developed profile provides invaluable information regarding the organization. A causal inquiry is a logical process of identifying reasons (causes) to explain the results of a specified event (in this case, the results of the organizational profile). Causal inquiry is part "connecting the dots" and part data interpretation. The evaluator combines his or her personal knowledge and experiences to determine if the organizational profile matches reality or is different. Often, reality (situational analysis and environmental scans) conflicts with the data obtained during the assessment phase. This results in further "what-if" analysis and the examination of alternatives.

When the data matches (or aligns), the next step is to review the available strategies to identify a strategy or modify an existing strategy. The project is ready for the N²OVATE™ process (discussed in a further chapter).

Summary

This chapter provided a preview of how diagnostics becomes an integral part of an innovation project. Analyzing the present situation and the environment created by and operating within the organization and combining it with assessment and interpretation enables an evaluator to formulate a "best-fit" strategy. Rather than trying to fit an innovation strategy to a project, our approach lets the process identify critical elements that work best with a selected strategy. Clients prefer a unique solution to a preformulated version. This also encourages buy-in and alignment, two key features of the N²OVATE™ methodology. Present the information in a summary (and descriptive) format to management, using the result for planning purposes. The better the information is, the less risk of failure exists. A further discussion on innovation strategies follows this chapter.

Discussion Questions

1. Discuss what skills evaluators need to conduct an environment scan.
2. What types of data are useful for evaluating opinions, feelings, and beliefs?
3. Describe a process of building a user (customer) profile.
4. Define, in your own terms, a causal inquiry.
5. How do your own personal experiences and knowledge prepare you to evaluate of an organization?

Assignments

1. Create a simple set of steps to complete a situational analysis.

2. Devise a simple process for conducting and then analyzing an environmental scan.
3. Compare and contrast passive and active data. What makes active data more valuable than passive data?
4. Build a simple profile from available information in your organization.

Chapter 8

Innovation Strategies: Design for Success

Introduction

For most agencies or organizations, innovations occur as a chance event (or are event driven). Typically, a critical need or requirement drives the organization to produce a product, service, or technology. Generally, the agency or organization assembles or forms a team or teams to address the need or requirement. The organization plans and directs activities focused on expending limited resources to achieve a benefit that far exceeds the overall cost. When innovation becomes a key objective for the organization, it can be difficult to plan for chance events or opportunities that often present the best cases for innovation. Further, often the difficulty is repeating the innovation success multiple times in a reactive environment. For sustained innovation success, we feel a proactive and determined strategy is required for any agency or organization postured to take advantage of potential innovation opportunities and discovery.

By our definition, a strategy is simply a plan to accomplish one or more objectives (goals) to meet a need or perhaps a set of requirements. To ensure sustained innovation success, agencies and organizations need a long-term plan that is consistent over time and resistant to frequent modifications and changes. The objectives must be measurable and the plan logical to succeed. The organization must commit to a long-term strategy combined with the desire to ensure that the innovation effort succeeds.

The next logical step after assessment and diagnostics is the development of an innovation management strategy. In simpler terms, innovation must be a corporate (executive) strategy, accepted and promoted by management, who operationalize it on a daily basis. Innovation is generally conducted on a project basis with two important but different aspects to characterize innovation. The first is identifying the innovation project (Chapter 9), and the second is implementing (managing) the innovation project (Chapter 10). The administrative component occurs at the executive (director) and level and above within the organization; the operational component functions as a project management (PM) strategy dedicated to achieving sustained innovation success.

Organizational Scale

Strategies for innovation for the federal sector can occur anywhere along the leadership structure or chain of command (see Figure 8.1). Agencies or organizations usually implement the innovation strategy, but the directive for innovation often comes from a higher authority. Strategies often begin at the national level, which is responsible for approving such initiatives. Certain entities within the government system have some level of discretionary spending allocated at the beginning of the year that they directly control. Here is where policy flows from requiring organizational or agency leadership and the chain of command to implement and succeed. Policy directives require information and interpretation at the next level

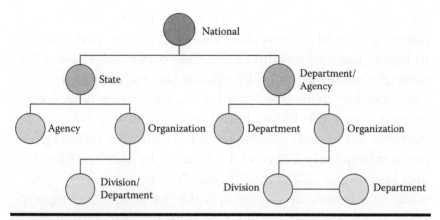

Figure 8.1 Organizational scale.

below. This continues until the operational level is reached (e.g., at the division or department level).

The interpretation requires thorough understanding of the mission, values, and purpose statement for developing a viable strategy. Smaller organizations may or may not have considered innovation as a strategic objective. In either case, what is often unknown is the type of strategy deployed. Knowing the type of innovation strategy deployed provides information on how to update or correct the strategy that permits and encourages the concept.

Existing Strategies

Many organizations subscribe to innovation and its benefits. For an effort, such as innovation, to sustain positive results requires a specific and focused strategy. Often, innovation is part of a greater organizational or agency strategy to promote value, efficiencies, and lower costs. Yet, effective innovation efforts require a separate, unique strategy and vision. Some of the more common existing innovation strategies are blue ocean, disruptive, and fast second, with a focus on achieving effectiveness as an overall strategy for innovation.

When companies create blue oceans, they break from their traditional strategy to infrequently pursue differentiation and low cost simultaneously (Buisson & Silberzahn, 2010). This limits the overall success innovation can deliver. According to Buisson and Silberzahn (2010), blue ocean and fast-second strategies simply do not fully explain successful innovation, and they maintained that innovative success is secured through four kinds of breakthroughs: design, technology, business model, and process breakthroughs. Breakthroughs occur infrequently and often by chance. In addition, Kim and Mauborgne (2009) provide that "by driving down costs while simultaneously driving up value for buyers, a company can achieve a leap in value for both itself and its customers" (p. 83); building new market segments requires a consistent pattern of strategic thinking, one that never uses the existing competition as a benchmark. These "leapfrog" innovation strategies may only come once or twice in the life of the organization. Missing are the benefits of change and incremental improvement. An example of the leapfrog innovation strategy is the iPod, which changed the face of the recorded music business. In fact, raising the social "cool factor" of your product line is definitely desirable; maintaining and sustaining that product line as more companies enter the market segment tends to narrow your profit margins as increasing supplies and lower demand drive down market share. Apple's iPod entry redefined the market standard and made other competitors retool or become irrelevant. As Apple looked for new opportunities to expand its spectrum of influence, the cellular handset industry was a new frontier waiting for a unifying champion—enter the iPhone.

Developed by Christensen (1997), the disruptive innovation theory challenges the traditional framework of sustainment strategies by altering the product or delivery methodology and displacing the existing process. Raynor (2011) adds that Christensen's theory can be employed to either enhance innovative concepts that currently exist or create balanced portfolios of innovation initiatives. The seminal principle in Christensen's theory is that companies innovate faster than the customer base demands. A good example of this is Apple's innovative i-products. Disruptive innovation can also have a negative side as it can cause disruptive effects within the organization. It is most useful on an organizational scale when product redefinition is required. Organizations that need to redefine themselves can find this strategy useful.

Perhaps a better example comes from the aircraft industry's leading provider, Boeing Aerospace Corporation. Sticking to a proven record of accomplishment of forging solid foundations before moving forward, Boeing has learned to listen to its customers while creating new and balanced portfolios that meet customer needs (Masters, 2007). With global business ethics gravitating more toward social responsibility, global partnering, and "greener" solutions, Boeing's new business model and integration of new aerospace structure technology (composite wings and fuselage), big sky interior, and global supply chain approach was fresh and destined for success. This large-scale global collaboration between Alena, Boeing, Fuji, Kawasaki, Mitsubishi, Spirit, and Vought at seven production sites around the world displays a diverse set of highly integrated partnerships that did not exist in the commercial aircraft industry. These new approaches broke virgin ground, and Boeing's new collaborative business model may become the road map for international collaboration and innovation effectiveness. Therefore, some innovation strategies follow a "home-grown" approach.

To examine internal innovation strategies, the next section examines different methods of categorizing innovation types. These simple descriptors will assist in evaluating the strategy's effectiveness, intent, and future sustainability.

Innovation Strategy Types

Classifying or categorizing the types of strategic innovation is critical to success and future growth. Clausen, Pohola,

Sapprasert, and Verspagen (2012) identified five unique strategies that agencies/organizations often employ:

1. Ad hoc: random innovations (strategies created on a project basis)
2. Supplier based: driven by newer technology, new products
3. Market driven: consumer (client or competitor) or user driven
4. Research and development (R&D) intensive: a traditional approach by which innovation emanates from one source
5. Science based: reliance on patents, science, and research

We have added a sixth strategy:

6. No strategy (author added): no discernible strategy or plan

The results of Clausen et al. (2012) support the idea that the differences in innovation strategies across firms are an important determinant of the firms' probability to repeat and sustain success. More concretely, they found that firms pursuing the market-driven, R&D-intensive, and science-based strategies were more likely to be persistent innovators and that persistence is as critical as a defined strategy.

To determine simplistically the strategy deployed requires some form of self-assessment. Use the following set of survey statements with a 5-point or 7-point Likert scale. The agreement/disagreement scale provides a method of establishing the prevailing strategy that an organization could deploy. Assign a value of 1 to strongly disagree and a value of 5 to strongly agree (using a 5-point scale; a 7-point scale would have more gradations, adding the somewhat disagree and somewhat agree values). Distribute to the director level and above (minimum of 20–30 participants). Calculate an average response for each group. Transfer the averages to a "radar chart" (Figure 8.2).

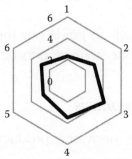

Figure 8.2 Radar chart: predominance of strategy type.

Strategy 1: Ad Hoc

1. Innovations are often unplanned.
2. Innovations are event based rather than strategic.
3. Innovation is a common workplace word.
4. It seems that innovations appear only when needed.
5. Innovation is discussed on a regular basis within the organization.

Strategy 2: Supplier Based

1. New technology drives the agency or organization to innovate.
2. When a supplier changes a part or product, our organization must innovate.
3. Our suppliers have a great deal of influence on what and how we innovate.
4. Working with suppliers is a way to determine our innovation needs.
5. We rely heavily on our suppliers to innovate.

Strategy 3: Directive Driven

1. Directives determine what and when we will innovate.
2. The innovation strategy is focused on new directives.

3. When the organization meets directives, innovation efforts can be successful.
4. We value user needs to guide our innovation strategy.
5. Directives guide our strategic efforts.

Strategy 4: R&D Emphasis

1. R&D creates most innovative products/services/technology.
2. R&D's primary function is innovation.
3. Administration and R&D set the strategic vision for innovation.
4. Innovation proceeds only with the approval of R&D.
5. The Engineering Department creates and manages innovation.

Strategy 5: Science Based

1. The agency/organization relies on patents to advance innovation.
2. A science and discovery focus is the result of the organization's innovation policy.
3. The agency/organization is research focused for its innovative products or services.
4. The agency/organization regularly advances the field of science and technology with its innovation.
5. A strong research focus continues to be the strategic vision for the organization.

Strategy 6: No Strategy

1. There is no specific or declared strategy associated with innovation.
2. There is a strategy but it lacks any relevant enforcement.
3. The strategy is weak and possibly not applicable.

4. No plans exist to implement an innovation strategy.
5. Innovation is a strategic goal but remains to be implemented.

For those organizations with a strategy, evaluate one of the first five strategy types. Analyze the averages of the 20 respondents to determine the predominance of innovation strategy type.

Example of 20 Respondents' (Director Level and Above) Averages

	Type 1	Type 2	Type 3	Type 4	Type 5
Average	2.3	3	4.1	3.5	2.7

The average points to the third strategy (directive driven) as dominant.

Many agencies or organizations have no strategic emphasis on innovation. A no strategy approach includes those agencies/organizations that give only lip service to innovative activities. Secretary-level executives can ask for an emphasis on innovation, but if the culture and environment do not work in its favor, the attempt to implement will fail. If the organization has failed to deliver successful innovations, then a no strategy type may be the best descriptor.

When analyzing the data, examine the range of strategy types selected. Variation is the choice of strategy type. If respondents choose more than one strategy type, this indicates lack of clarity, poor follow-up, or communication difficulties. Multiple choices for a predominant innovation strategy suggest that no clear strategy exists. Use the results to drive a discussion on modifying or creating a viable strategy. For those who "pass" with good agreement, consider the organization ready for a more detailed and specific strategy.

For those who experience difficulties, consider an additional step. Examine the mission, vision, and purpose statements of the existing strategy.

Mission, Vision, and Purpose Statements

To understand and implement the strategic component, the agency or organization must first examine its mission, purpose, and value statements. These statements provide a view of what is important to the entity, the environment that innovation must thrive in, and commitments made on activities such as innovation. These statements provide the reason for existence and the goals and aspirations of the organization. The choice of words is critical because these uniquely describe the organization and its intent. Not only the choice of words is important but also the arrangement and emphasis of the language as key elements contribute to the overriding mission of the organization.

Innovation may often be an outcome of the mission, vision, or purpose statement. If so, then the words that describe how the organization will achieve such a goal are critical. Therefore, examining these statements provides an insight into how the organization attempts to achieve these goals. For the federal sector, often the mission, vision, and purpose statements begin at the national level. Agencies and organizations that develop strategies must adapt to all national directives. Innovation is a common directive for many agencies. These agencies often create similar descriptive statements: innovation as a key strategy.

A simple method of assessing the innovation strategy of any organization is to examine the language of these statements. Each agency must search for key words linked to innovation success. Some common terms are given in Table 8.1.

Innovation cannot thrive in an organization committed to unrelated goals or objectives. Therefore, innovation may begin by correcting or updating the organization's purpose for existence. Expecting innovation success from an agency dedicated to some other goal or objective is ludicrous. Yet, language exists to recognize innovation without any effective

Table 8.1 Key Words Supporting an Innovation Strategy

Success Measures	Environment	Commitment
Reduced costs	Cooperative	Matching resources
Efficiencies/effectiveness	Creative	Long term
Added value	Recognition and rewards	Dedicated personnel
High performance	Trust	Leadership

support mechanisms. First, examine these strategic descriptors of purpose and objectives. Review to determine if these documents have a pro, neutral, or negative slant to innovation. If a conflict exists with innovation descriptors, chances are innovation will not succeed because there is no directive to meet this objective. If the mission, purpose, or vision statements do not contain this language, then the strategic elements may not exist to support innovation. If innovation is present, then these descriptors may send a mixed signal.

Financial results are not enough of a driver for innovation. The environment that supports creativity, cooperation, and coordination must be present if all are to participate. Otherwise, innovation may be limited in scope and benefit. Finally, management must lead the effort by committing resources and personnel for success. Changing the mindset to one more open to innovation may be difficult to accomplish and time consuming and may upset many. Those organizations without a corporate vision that promotes innovation will need to start at this very step.

Most large agencies or organizations have developed a unique culture. This culture may come from its past, its mission, or its directives. For example, the Central Intelligence Agency (CIA) has a unique culture developed to support its ever-changing mission. Technology is such an integral part of the CIA's modus operandi that innovation is a natural part of its mission and existence. Another natural is the National

Aeronautics and Space Administration (NASA), which continually produces and releases new products and new technology. It is almost an acronym for innovation. What about agencies such as the Internal Revenue Service (IRS)? Its focus is on collection of taxes and fees. Does this lend itself to innovation, or will its mission and purpose require an adjustment or a culture change to operationalize innovation?

If descriptions of the culture do not include words such as cooperative, collaborative, or challenging, then innovation is of secondary importance. That culture, which has developed over time, is the result of a specific strategy deployment. If innovation is not part of that culture, then it will be difficult to deploy. Begin by changing the culture with a strategy that supports innovation success for the long term (discussed in more depth in the next chapter). Remember that success breeds success. Use a proven methodology to demonstrate success at the project level. Use these successes (with both outcomes and personnel) to develop the fundamentals of a strategy.

Smaller organizations may or may not have considered innovation as a strategic objective. For these organizations, the focus will be on innovation projects that result in incremental improvements as well as large-scale projects. For innovation to successfully "pay back," the organization needs a comprehensive methodology and strong management leadership and support.

Many agencies or organizations have no strategic emphasis on innovation. A no strategy approach includes those agencies/organizations that only give lip service to innovative activities. Setting a goal (such as innovation) and then providing no plan to achieve the goal produces little value. Secretary-level executives can ask for an emphasis on innovation, but if the culture and environment do not work in its favor, the attempt to implement will fail. Without a systemwide strategy, innovation exists as a random and uncontrolled occurrence. Planning for these events is more chance than choice.

Compatibility with Organizational Outcomes and Functions

Compatibility is an issue when aligning an innovation strategy with organizational structures. Reasons for this disconnect include the complexity of the innovation attempted and the type of organizational structure selected. Innovation complexity ranges from incremental improvements up to radical changes, such as in disruptive innovations, discoveries, or significant patents. The more complex the innovation, the more heavily invested the organization will need to be. Both the amount of resources and the threat of risk increase with each subsequent innovation stage. Contrasting the innovation complexity with organizational functions, systems, and management creates nine unique innovation strategies profiled in Figure 8.3. Those highlighted in gray are under development at this time.

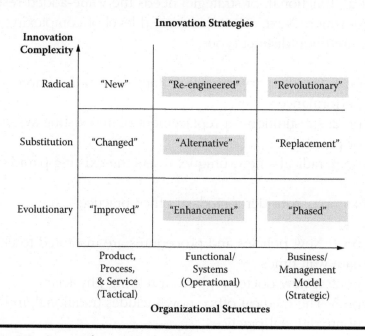

Figure 8.3   Innovation strategic grid.

Agencies and organizations are searching for strategies that will produce sustained success. Leaders and those organizational decision makers must understand that the intent and form of innovation are different depending on its application in an organization. The innovation strategy needed to improve the decision-making process is different from improving the performance of a product or service provided to users. Trying a "one-size-fits-all" approach will be both frustrating and nonproductive.

From our perspective, different outcomes and applications require different innovation strategies. Given that, customers (users) want innovative outcomes; this was the first strategy we developed. The strategies presented in this book address process, product (technology), and service applications. The innovation strategy should complement the overall organizational structure. To examine for discrepancies and identify opportunities, first determine which organizational structure (tactical, functional, or strategic) needs the value-added results of innovation. Next, select the desired level of complexity; there are three distinct types:

Type 1: evolutionary—ongoing improvement for improved performance
Type 2: substitution—a replacement of an existing system or practice
Type 3: radical—new, unique, resets the existing paradigm

For example, consider a medical office practice:

Type 1: New policies and procedures are instituted to eliminate wait times.
Type 2: A new doctor replaces a retiring physician.
Type 3: The medical office now includes traditional and nontraditional practices.

The agency or organization must decide on the most appropriate choice. On the horizontal axis are three organizational structures. The structure refers to where the innovation strategy is applied. There are three categories of structure:

1. Product, process, and service: external (user) focus (tactical)
2. Systems: internal operations, practices, policies, and procedures (an internal focus) (functional/operational)
3. Strategic: leadership and decision making

It is common to want innovation to occur across all three structures. Realize, though, that the strategies could be different.

The internal grid are key words that suggest the appropriate strategies to employ. The organization could determine that the incremental approach works best for all three functions. Yet, for the business management model, the approach is different. Different functions may require modified strategies. For example, an innovation strategy for Human Resources is not compatible with one for the Accounts/Finance Department, although there are many common elements. Organizations may want three separate strategies for their external focus product, process, or service functions, yet want only one strategy for their internal systems and business model. The N²OVATE™ model (with modifications) will work successfully in all nine strategies.

Creating individual strategies for each element in the grid is beyond the scope of this book. This text discusses the strategies for innovating product, process, and service across all three dimensions of innovations (Chapters 9 and 10). Look for future books to develop these additional strategies for successful sustained innovation.

Constructing an Innovation Strategy

Rather than constructing a new set of values, mission, or purpose statements, begin first with an assessment of internal and external capabilities. Perform a SWOT analysis: S for strengths, W for weaknesses, O for opportunities, and T for threats).

Table 8.2 List of Characteristics Considered When Performing a SWOT Analysis

Characteristic	Positive/Negative
Financial	Profits, net income, ROI, ROA, ROE, costs, taxes, fees, etc.
Strategic	Marketplace standing, number of competitors, competitive advantage, product, service or technology competency, history, product line, product life cycle
Resources	Availability, cost, supply chain, integration, outsourcing, staff competency, storage, global capacity
Knowledge	History, accountability, learning culture, technical competency, inherent versus learned, creativity
Operational issues	Staffing, capability, delivery, maintenance, training, global response, turnaround time, business processes
Long versus short term	5-year planning horizon, immediate needs, enterprise solutions
Standards, quality, reliability	Measures, best practices, quality practices, warranty, guarantees, etc.
Customer/user relationships	Satisfaction issues, consumer behaviors, complaints, effective response

Note: ROA = Return on Assets; ROE = Return on Equity; ROI = Return on Investment

Table 8.2 describes some characteristics to consider when determining the elements that constitute a SWOT analysis. This list is not exclusive, but it does consider the issues when attempting a SWOT analysis (Bevanda & Turk, 2011).

Strengths, Weaknesses, Opportunities, and Threats Analysis

A SWOT analysis is useful for developing the elements of an innovation strategy. Consider a concept, such as innovation.

How can an organization design an innovation strategy to maximize its strengths (opportunities) and minimize its weaknesses (threats)? A SWOT analysis is an excellent tool for assessing these four characteristics. The SWOT characteristics are useful for developing a mission and value statement. This is how an innovation strategy begins. The next steps in constructing a strategy consist of developing a process, identifying outcomes, constructing control metrics, assigning responsibilities, and performing validation.

Gather a team of executives, directors, and managers. For each category (e.g., strengths), brainstorm a list of characteristics (Table 8.2) of the organization that are defined strengths. Brainstorming is a process, not a chaotic generation of ideas. Brainstorming consists of topic selection, individual idea generation, writing (list of ideas or concepts), and evaluation. The result of the exercise is a list of items related to the topic. Further evaluation can consist of ranking the ideas/concepts.

Begin with strengths; consider the questions in Figure 8.4, then do the same for the remaining three categories. Be

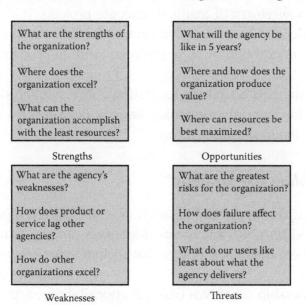

Figure 8.4   Questions to answer when completing a SWOT analysis.

prepared for discussion and negotiation. The primary intent is innovation. For every strength, there is a potential weakness (Evans & Wright, 2009). Address threats (both internal and external). The four quadrants will help develop a sound strategic innovation policy.

Ghazinoory, Abdi, and Azadegan-Mehr (2011) report that using SWOT analysis to help develop an innovation strategy is unique. Yet, it follows that a well-managed business or organization will be more successful. The innovation strategy should build on strengths and opportunities while considering the risks, threats, and weaknesses. Be aware that a weakness may already be visible to your customer or user.

The organization can construct well-developed mission, value, and purpose statements; create metrics to measure innovation progress; and manage the process to produce acceptable results. Additional critical elements of a successful strategy are as follows:

1. Communication of message and intent
2. Empowerment of teams to execute projects
3. Leadership to guide execution and follow-through
4. Secure embedding in department and division policy

What remains is a strategy for lasting and results-producing innovation. The only remaining issue is persistence—what to do to maintain the momentum.

Assembling the Strategy

After completing the SWOT analysis, establish the mission, value, and purpose statements. Identify a process to meet the outcomes of these descriptors that includes the following:

1. Leadership directives (roles, responsibilities, actions)
2. Degree and intensity of support

3. Measurement and evaluation to assess performance
4. Review, modification, and replacement policies

Implement at a planned pace. Try with one division or department before implementing organizationwide. Review and evaluate; be willing to change and modify for the most successful outcome. Once tested, roll out to the entire organization. Be keenly aware of persistence issues.

Persistence

Persistence is a critical element of any innovation strategy. Even the best ideas fail at times to achieve a desired performance level or meet specific needs. Humans learn more through failure than success. Given this advice, an organization's innovation strategy (and efforts) must prevail in the face of occasional failures. Persistence is more than just innovating repeatedly; it is also a mindset. Pharmaceutical organizations that innovate by discovering or inventing a new drug may fail repeatedly before achieving success. Persistence is a corporate or organizational trait that resists failure but "learns" from its mistakes. Include measures that track persistence, consistency, and lessons learned.

Implementing an Innovation Strategy

Finally, the remaining item is to implement an innovation strategy. As with any strategy, one size does not fit all, meaning that a single strategy is not applicable throughout the organization. Often, organizations determine that a strategy conceived (designed) at the executive level will apply to the entire organization. These strategies have a short shelf life given their inability to adapt to various corporate, operational, and user-required requirements and situations. Recognizing the need for an adaptive strategy is the first step in developing sustained innovation success.

Strategies need refinement and require periodic adaptation. Figure 8.3 highlights nine unique grids that require an innovation strategy. Of course, much of the process is transferrable, but each grid requires a unique combination of tools, resources, management commitment, and support. Consider these five competencies as the organization develops its adaptive innovation strategy:

1. Involvement (leadership, managerial, employee)
2. Proven success rate (number of projects, success/failure rate, risk)
3. Employee and user awareness (communication effectiveness, feedback, idea generation)
4. Commitment (determination of overall support)
5. Resources (ability to execute without hardships)

Developing these strategies is not complex but requires some combination of all five competencies. Avoiding one or more is an invitation for failure.

Developing a basic strategy requires the use of the tools described in this chapter. After aligning the values, mission, and purpose statements, develop a SWOT analysis to determine elements of the strategy. Create a process flow diagram and identify critical junctures and conflict points. Next, check the strategy against the five competencies for compliance or potential vulnerabilities. Have a team review the strategy before implementation. Create evaluation "points" to ensure accomplishment. Conduct a review session after the first 3–6 months. Modify what is not performing well. Follow up with yearly reviews.

Strategies will build and evolve over time. A final key element of a strategy is the evaluation of outcomes (results). Innovation is not magic and will not always generate spectacular results or a cure for every problem. It will follow a set method, and the results should satisfy an unfulfilled need in some form. In essence, innovations should meet certain (and identifiable) factors that define whether the innovation is productive.

Success Factors

For innovation to succeed, it must meet a set of success factors that define value to both the organization and the user. These success factors define whether the innovation project (or strategy) accomplishes its objectives. Success factors come from the organization, its leadership, and the customer (user) and measure accomplishment. Determining these success measures or factors enables the organization to measure progress, highlight problems, or identify opportunities. The user and the organization meet their objectives and goals by generating these metrics. Each project or each department will have a different (but related) strategy. These success factors become "dashboards" or "scorecards" for the entire innovation effort (Figure 8.5). By monitoring

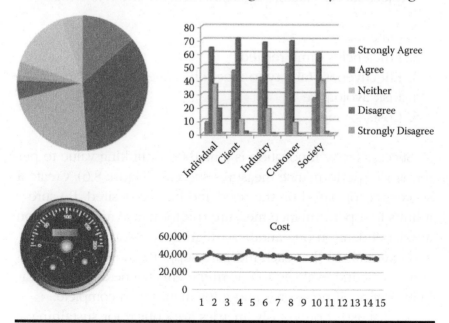

Figure 8.5 Typical dashboard (scorecard).

these dashboards, management can easily identify problems or opportunities.

Success factors define a measure (metric) tied directly to an objective or goal. These dashboards work for both an individual project and overall innovation strategy monitoring. The focus is on outcomes and then the process necessary to achieve these outcomes on a recurring basis (sustained success).

From a strategic perspective, the success factors will evaluate the "health" of the innovation effort. Typical strategic success factors will measure items such as the following:

1. Number of successful projects implemented (completed)
2. ROI (return on investment): more benefit from the innovative approaches
3. Value generation
4. Customer (user) acceptance
5. Quality or satisfaction

Project success factors typically measure the following:

1. Outcomes
2. Efficiencies (time related)
3. Effectiveness (ability to meet objectives and goals)
4. Risk avoidance
5. Costs/savings

Success factor innovation is a method of linking value to performance (performance measures success) (Figure 8.6). Create a success factor based on the need and benefit desired. Be sure it links to a performance measure (performance and innovation are dependent on one another). Align these measures (internal and external) so that each measure supports the intended outcome. Test and evaluate to determine effectiveness and applicability. Refine, if necessary, and implement when complete.

As discussed previously, metrics are critical for sustaining a strategy. Success factors measure performance, progress, and

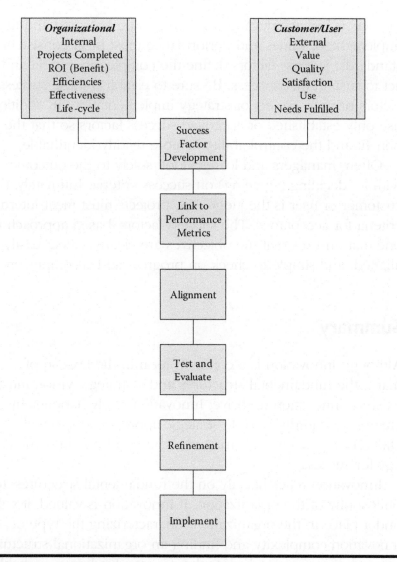

Figure 8.6 Success factors.

effectiveness. These measurements act like a set of competencies or factors that define success. As part of a coordinated strategy, meeting these factors will determine project outcome success. If an innovation project (effort) meets these factors, it has achieved its objective, met the appropriate needs, and provided benefit as defined by the organization and its leadership and customers (users). Identifying success factors for smaller projects is just as critical. If these are to be accepted and then implemented, values and performance must meet consistent standards. Success factors define the completion criteria and act to sustain that success. Be sure to establish these success factors prior to project or strategy implementation. In addition, use only established or approved success factors so that the benefit and the control ability is more readily identifiable.

Often, managers and leaders look solely to the outcome without deciding (agreeing) on success criteria. Ultimately, the customer or user is the judge. Yet, projects must meet internal criteria for acceptance. The success factors–based approach is one that ensures that the objectives are clearly stated, easily aligned, and simple to check on progress and accomplishment.

Summary

Although innovation is a common term in the lexicon of many, the fundamental structures and a strategic vision must exist for innovation to thrive. Innovation rarely happens by chance; it requires a viable strategic approach to succeed. Different organizational structures will require a different strategy for success.

Innovation relies heavily on the fundamental structures and philosophy of the organization. If innovation is valued, it will find a place in the organization. Characterizing the type of innovation complexity and finding an organizational structure in need of innovation begins the process. Finding a workable strategy is the key to sustained success.

Characterizing that strategy was one of the key objectives of this chapter as well as developing or evolving a present strategy to ensure continued innovation success. We choose the most common functions (with an external focus) in which to develop the innovation strategy described in the next two chapters.

Discussion Questions

1. Do the mission, values, and purpose statements indicate whether the organization or agency is innovative?
2. Is it enough for an agency or organization to be committed to innovation? If not, on what path would you suggest the organization embark?
3. What are the advantages and disadvantages in conducting a SWOT analysis before developing an innovation strategy?
4. Use Figure 8.4 to determine which organizational function requires the application of an innovation strategy. Is it possible to obtain satisfactory results by using a simpler innovation approach?
5. Describe one organizational success factor and the method used to construct, test, and verify that measure.

Assignments

1. Evaluate your organization using the strategy type indicator. Select a colleague or friend to do the same. Examine the differences and similarities and summarize the results. Do you both agree on the predominant culture trait?
2. Do a simple SWOT analysis on a process, product, or service with which you are familiar. Which quadrant predominates, and what is needed to innovate that product, process, or service?
3. Develop a set of success factors for a process, product, or service.

Chapter 9

Selecting an Innovation Project: Projects that Add Lasting Value

Introduction

Once the corporate strategy is underway (or in development), a tactical strategy can begin full implementation. In this chapter, the focus is on the development and application of a tactical innovation process for the federal sector. *Tactical* refers to more of a day-to-day product, process, and service application. The focus is on the external primarily as this is the touch point where the customer or user determines the value of the innovation. The information presented in this chapter is for a generic process, one that requires improvement in some form. Our practice is to develop a highly customized set of strategies for each agency or organization. To demonstrate the process, the focus is on improving an existing process.

Due to its unique nature, we have developed a methodology specifically for the federal sector. Because innovation usually involves a project-based approach, this chapter highlights a process that validates the best projects for implementation.

Project validation is a key tactical strategy critical for individual and ongoing innovation project success.

It is an opportunity to present the N²OVATE™ methodology, which stands for

Step 1: N = Needs and new ideas
Step 2: N = Normalize and nominate
Step 3: O = Objectives and operationalize
Step 4: V = Verify and validate
Step 5: A = Adapt and align
Step 6: T = Tabulate and track performance
Step 7: E = Evaluate and execute

This next-generation methodology is built on the ENOVALE™ framework developed in *Chance or Choice: Unlocking Sustained Innovation Success* (McLaughlin & Caraballo, 2013a). ENOVALE™ was developed for typical business applications. N²OVATE™ fits the government-sector profile.

To present this methodology, a hypothetical case study using all required steps is offered. The case study follows a project from idea generation to acceptance or rejection. It is presented to better explain the methodology. Remember, this initial phase develops and validates an innovation project that can produce true value. Actual implementation is discussed in the next chapter.

The case study begins, as always, with a need and at least one objective that has quantifiable value. The need is critical but unsatisfied. For this example, the need is for a more effective (efficient) aircraft parts procurement system to replace an existing system initialized in the early 2000s. These parts are specifically for the F-22 Raptor fighter aircraft. This system does not include avionics, defensive, or communication systems. The parts associated with this case are not high risk and are classified (e.g., require a high security clearance) but are used for general maintenance. One objective is that the parts arrive on time; another is that parts meet standards; a third is to control costs. The objectives are the innovation success measures.

Step 1: Assessment

Data analysis has shown that simple high-wear parts replaced during the required A checks* (a light, routine maintenance and operational systems inspection that occurs every 50 flight hours, e.g.) or under an established preventive maintenance cycle that frequently takes a lead time of 60–90 days to acquire from supply chain sources. This requires additional inventory, which is an added expense. The need exists for a more efficient and effective procurement system (the objective). Figure 9.1 provides a process flowchart of the needs generation as it relates to the objective. As part of the process of assessing the need, Figure 9.2 identifies the innovation type that must meet three criteria (viability, capability, and sustainability). Viability refers to the usability of the item; capability indicates whether it will meet objectives consistently; and sustainability refers to the life cycle of the item. For this procurement system case study example, the three criteria of needs are as follows:

Viability: Changes would improve performance and reliability.
Capability: The system must be cost effective and function within stated parameters.
Sustainability: The procurement system must be valid and timely and add more value than the previous process.

---

* A checks are normally conducted every 500 flight hours (for heavies [large, wide-body aircraft], but it varies on the airframe or type of aircraft) per civil aviation authority directives. B checks are more extensive, and this also varies by airframe. A and B checks can happen overnight virtually anywhere, but C and D checks are more extensive and require the aircraft to be in a hangar for a period of time. C checks are performed every 12–18 months and D checks (most extensive) every 4–5 years. *Air Force One*, for example, *could* feasibly go through a complete D check every two years for obvious reasons (a modification to the typical civilian aviation authority requirements).

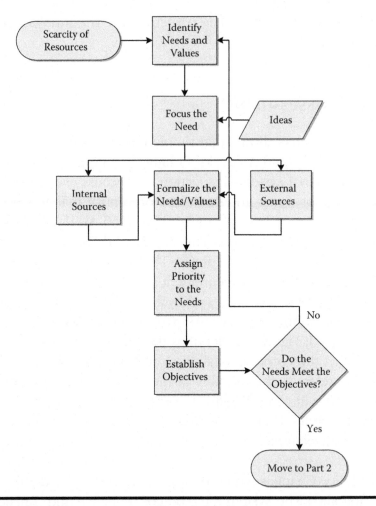

Figure 9.1 Step 1 (part 1): needs and new values.

If the criteria are acceptable and meet the overall objectives, the project can proceed.

Step 2: Normalize and Nominate

Once the need is established and verified, the next step can proceed. Before forming the team, assess individuals on their innovation perceptions, work environment perceptions, and what each person values. Have the team complete the simple

Selecting an Innovation Project ■ 171

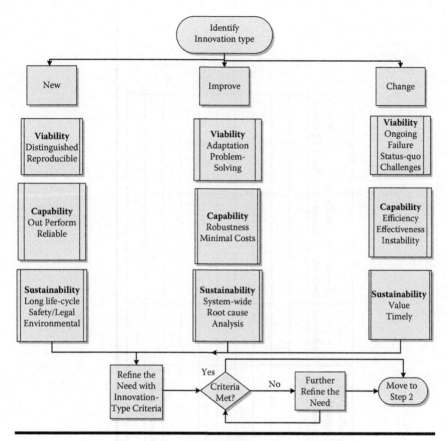

Figure 9.2 Step 1(part 2): needs analysis.

10-question survey seen in Figure 9.3 to begin the normalize-and-nominate step (Figure 9.4). Assign a numerical score to each response: 1, undefined; 2, poorly defined; 3, minimally defined; 4, defined; or 5, well defined. Calculate an average and range of the following descriptions:

New Average: (Statements 1 + 2 + 8)/3;
Range Max–Min =

Improve Average (Statements 3 + 5 + 6 + 9)/4;
Range Max–Min =

Change: Average (Statements 4 + 7 + 10)/3;
Range Max–Min =

Statement Number	Instructions: Check the Box that Best Matches Your Understanding of How Innovation Is Defined by Each Statement	Undefined	Poorly Defined	Minimally Defined	Defined	Well Defined
1	A new discovery or invention					
2	New or novel (unique) ideas					
3	Making something better					
4	Replacing what does not seem to work					
5	Continuous improvement					
6	Improving something to make it better					
7	Changing what does not work					
8	A new invention or patent					
9	Improving on something that already exists					
10	Changing for the better					

Figure 9.3 Innovation comprehension survey.

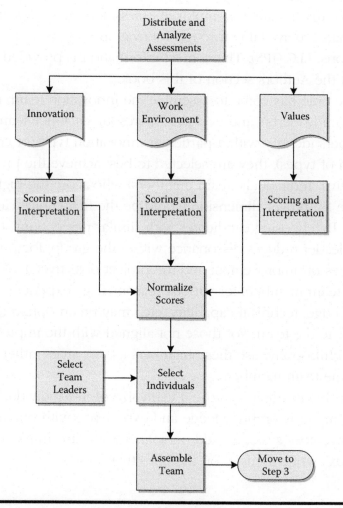

Figure 9.4 Step 2: normalize and nominate.

Whichever of the three components of innovation comprehension scores the highest is the preferred method of recognizing innovation. In other words, you are looking for a range value that is less than 1.5. A higher value indicates mixed feelings or opinions. Responses from this small survey are not conclusive and will provide an understanding of what the respondent perceives (understands) about innovation. Like individuals, those who share similar sentiments regarding innovation will work best on a team.

Note that additional employee assessment resources and materials are available through Innovation Process and Solutions, LLC (IPS). The contact information is provided in the About the Authors section of this book.

The time has come to assemble the innovation team, name the team leader(s), and evaluate the resources. Once team members identify with a particular innovation type (or combination of types), they are selected to best achieve the project objective. Individuals (team members) who score the highest on the "improve" dimension would be the first-choice candidates. By choosing candidates with similar perceptions, the team leader reduces dissonance within the group, focusing energies on improvement and the project objective(s). Adding other team members for various reasons (e.g., experience, knowledge, technical capability, etc.) may be an option to complete the team. For those not aligned with the improvement philosophy, use these members as resources rather than full-time team members.

For this example, choose a team size of 5–7 individuals with emphasis on knowledge and experience with purchasing and inventory systems. Add personnel who can "think outside the box" while focused on the objective.

Step 3: Objectify and Operationalize

Figure 9.5 displays the objectify step used to develop and describe the objective. Most projects have multiple objectives (values) that are focused on use, purpose, and intent. To develop an objective, use the SMART (specific, measurable, achievable, relevant, and time-based) criteria. For this example, there are three SMART objectives, based on a set of values, for the procurement process:

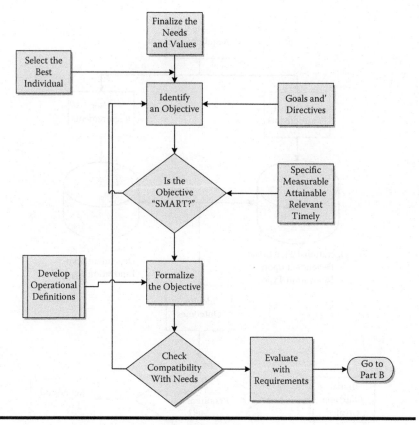

Figure 9.5 Step 3 (part 1): objectify step.

1. In place and operational within 60 days
2. Cost effective
3. Flexible and reliable

All of these descriptions (of the objective) require an operational definition (Figure 9.6).

An operational definition is one that describes the word and identifies parameters and measures. The word *cost effective* requires an operational definition. For this project, a cost-effective procurement system (process) is one that is, at minimum, capable of operating at a cost of 10 percent less (maintenance, personnel, distribution, warehousing, etc.); that is, the total monthly (quarterly) costs are 10 percent less than the previous

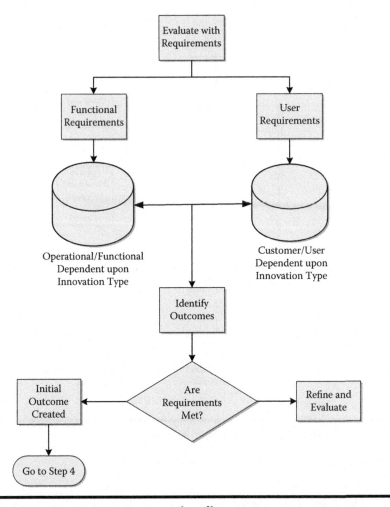

Figure 9.6 Step 3 (part 2): operationalize step.

version. Similar definitions for *flexibility* and *reliability* must occur. Often, this is one of the first tasks of the team. For cost-effectiveness, the objective is specific, measurable, attainable, relevant, and time bound (SMART). Poor or inaccurate measures will have a major effect on project success.

Finally, verify that the defined objectives are compatible with the need. Obviously, the procurement system must be a dramatic improvement over the existing process to satisfy the need. If not, then return to searching for the system that meets the needs of the agency or organization.

Table 9.1 Detail of Process (Item) Requirements

User or Functional	Requirements	Established Parameters
Functional	Integrated	Interfaces with existing systems and hardware
User	Simple to operate and navigate	Minimum training required Windows driven Accessible help screens
Functional	Easy to track	Parts tracked by purchase date
Functional	Reliability indicator	Tracks time between failures Estimates failure rates
User	Maintainable	Internal information technology personnel certified

Once the team finalizes the objective, operationalize the objective by defining the requirements (both functional and user) (Table 9.1). Here, the word *requirements* relates to operational parameters. At the project selection phase, these can be high-level requirements.

However, before the project begins, develop a more explicit set of requirements. One additional challenge is to understand the relationship between user and functional requirements. Table 9.2 provides a mechanism for evaluating these relationships. Strong or medium-size relationships require full exploration (Table 9.2). A relationship exists when functional and user characteristics affect one another. That is, as the functional requirement changes, it affects the user. The strength of these relationships may directly influence the objective.

For example, consider a case study (Table 9.2) for which functional and user requirements are an issue. The team wants to evaluate how three-success factors relate to selected functional concerns of interest. Assume that each person on the team decides to identify a relationship from his or her

Table 9.2 Functional–User Relationship Matrix

Functional	Customer or User Characteristics		
	Maintainable	Navigate	Ease of Use
Reliable	S	S	S
Simple to track	S	M	S
Interfaces	M	M	S
History	W	W	M
Detail	S	S	N
Comprehensive	S	S	S
Adaptable	S	M	M
Auditable	W	W	S
Accuracy	S	S	M
Startup	M	S	S

Note: Assign priorities: S = strong relationship; M = medium relationship; W = weak relationship; N = negative relationship.

perspective. This matrix provides a method of examining relationships from a basic, personal perspective.

Finally, the outcome (the objective matched with its requirements) is established. If there is a mismatch, the potential project is highly questionable. For the case study example, if the system is not fully integrated, simple to use, and easily maintainable, then the project may not satisfy its objectives. If the team reaches an unsatisfactory conclusion, then the project may require further refinement or be incompatible with stated objectives. If the assumption is that the project meets its objectives, the team can proceed to the next step.

Step 4: Validate and Verify

Once the outcome is accepted, it is time to validate and verify. Making a rash decision based solely on an outcome may

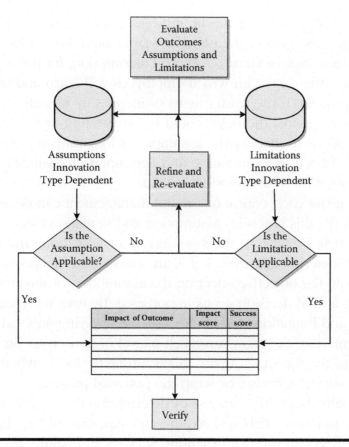

Figure 9.7　Step 4 (part 1): validate step.

eliminate potential projects while increasing risk and failure. It is best to enter into this step and determine if the outcome is worth the overall investment of time, resources, and effort. The beginning stages of this step require an evaluation of major assumptions and limitations (Figure 9.7). "A limitation is that which prevents, deters, or interferes with performance" (McLaughlin & Caraballo, 2013, p. 103). Limitations restrict performance and the ability to maximize effort. A potential limitation for this example is that the original system works but is based on the year 2000 technology. Potentially, even the most robust systems may encounter integration delays.

An assumption is intangible, based on available information, perception, fact, and opinion. Future performance will validate all assumptions. The more assumptions the project requires, the greater the need for viable alternatives given the higher risk of failure. A typical assumption, for this case study, is the ease with which employees will learn and use the new process. If the team misses or incorrectly identifies critical assumptions, the delay could be devastating or costly. The team needs to identify the assumptions for evaluative purposes. Check each limitation and assumption for validity (that the limitation or assumption is valid).

Evaluate each critical (team and management can decide which fits this criterion) assumption and limitation (see Table 9.3). Detail how the assumptions or limitations are measured (determined), even if it is an opinion or perception. Evaluate the objective effect on the assumption or limitation. Reach a final decision on its importance. By reviewing assumptions and limitations (use Table 9.4) and applying an evaluative criterion, the team can reach a level of consensus on the state of the outcome. Again, this is a time to decide whether to move forward, revise, or scrap the potential project.

Finally, begin the process of determining the overall impact of the outcome. This will require an estimation of the chance of success and the risk of failure. Success and failure are not mutually exclusive—one aspect can succeed, another fail. This is why the team receives training to use such tools to maximize the chance of success. Suppose the team arrived at three outcomes. These are listed in Table 9.5. In addition, the team evaluates for overall impact and success using a rating scale of 1 to 5. The team can convert this rating scale to a chart to identify best choices (see Figure 9.8).

1. Low or no impact, very small chance of success
2. Minimal impact, small chance of success
3. Moderate impact, medium chance of success
4. High impact, good chance of success
5. Superior impact, excellent chance of success

Table 9.3 Evaluation of Assumptions/Limitations

Assumption/Limitation	How It Is Measured	Effect on 60-Day Timeline	Effect on Cost Effectiveness	Effect on Objective Flexibility and Reliability	Overall Importance
L: Older technology	Date of inception	Strong	Strong	Moderate	Critical
A: Ease of use	Perception/mistakes and failures	Strong	Strong	Strong	Critical

Table 9.4 Effects and Importance of Assumptions and Limitations

Determinants of Effect	Measures of Importance
Strong	Critical
Moderate	Moderate importance
Weak	Unimportant

Table 9.5 Impact versus Success Evaluation

Impact of Outcome	Overall Impact	Chance of Success
Customized solution	1	5
Off-the-shelf purchase	3	3
Upgrade of software/ hardware	4	4

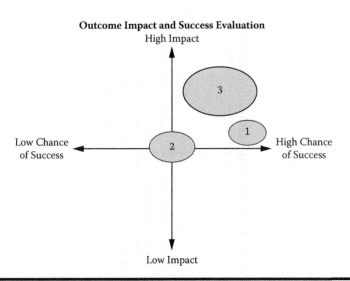

Figure 9.8 Outcome impact and success evaluation.

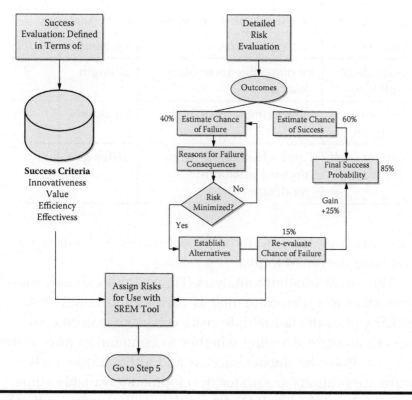

Figure 9.9 Step 4 (part 2): continuation of validate step.

Continuing the validation stage requires further verification. Validating outcomes requires a detailed risk evaluation. Evaluating risk takes on a complexity that is not addressed in this book. Yet, to evaluate a project, evaluate risk from a basic approach. Risk accumulates as the potential for failure (reduced performance) increases. One excellent method is the use of failure modes and effects analysis (FMEA) described in *ENOVALE™: How to Unlock Sustained Innovation Project Success* (McLaughlin & Caraballo, 2013b). Another, simpler approach is to assign a probability of success and failure, as shown in Figure 9.9. Finally, an alternative approach is to create a SREM (success, risk evaluation matrix) diagram (see Table 9.6). Locate the particulars on this technique in *Chance or Choice* (McLaughlin & Caraballo, 2013). Extract those

Table 9.6 Success, Risk Evaluation Matrix (SREM) Analysis

Success	Risk	Quadrants	Number
Upgrade of software	Incompatible technology: low	Strength	2
Customized solution	Time to implement: high	Weakness	4
Off-the-shelf software	Cannot handle special or unusual situations: moderate	Status quo	3

successful outcomes from the impact success evaluation and complete the SREM format.

The SREM quadrant analysis (Figure 9.10) exposes success measures (determinants) to various and critical risks. SREM can evaluate multiple risks versus an outcome success to assist in deciding whether to continue to pursue the project. Both the impact/success and SREM model software applications are under development (available some time in 2016). Figure 9.11 details the preliminary results of

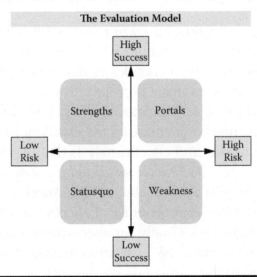

Figure 9.10 Success, risk evaluation matrix (SREM) quadrants.

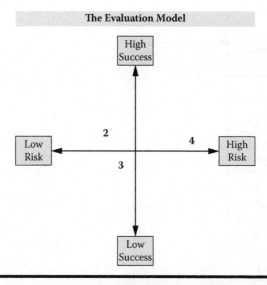

Figure 9.11 Success, risk evaluation matrix (SREM) project analysis.

the analysis, suggesting that the upgrade of software and hardware may be the best solution given the three unique objectives.

Step 5: Adaptation and Alignment

On accepting the outcome, now referred to as the innovative outcome, the team examines the human impact of the proposed project. Examining the human impact involves a determination of how well the project aligns with the team and organizational goals and values. Alignment begins with the ability to associate the outcomes with organizational values, culture, and scope (Figure 9.12). Overall, attaining value for the organization and user is the ultimate prize. If the project can demonstrate cost savings and improved efficiencies over time and value, it is a project worth pursuing.

One efficient method to measure alignment is through perceptual surveys, interviews, or focus groups. This includes an evaluation of both expectations and perceptions (Figure 9.13).

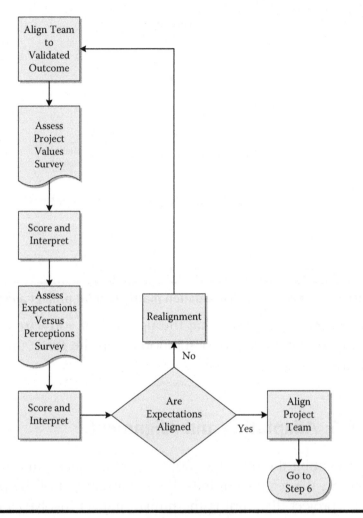

Figure 9.12 Step 5: alignment and adaptation.

For the scoring guide, there are five pairs of data (1, 2), (3, 4), (5, 6), (7, 8), (9, 10), and five concepts: project objectives, project performance, management contribution, team performance, and project outcomes (Figure 9.13). Assign a numerical value as follows: 1, strongly disagree; 2, disagree; 3, neither disagree nor agree; 4, agree; 5, strongly agree. Subtract the even-numbered statement from the odd-numbered statement. For example, select the project objective statements (statements 1, 2 from Figure 9.13). If an individual scores statement

Statement Number	Instructions: Check the Box that Best Matches Your Agreement or Disagreement with Each Statement	Strongly Disagree	Disagree	Neither Disagree nor Agree	Agree	Strongly Agree
1	I expect the project will achieve its objectives.					
2	In situations such as these, I have seen that projects will meet their objectives.					
3	This project should perform as expected.					
4	My experience tells me that projects like this one will perform.					
5	I expect management to contribute to project success.					
6	To date, management has continued to support process success.					
7	I expect that the team will exceed expected performance.					
8	The team has performed throughout this project.					
9	I expect few, if any, changes to the project outcome.					
10	The project has remained relatively unchanged throughout this process.					

Figure 9.13 Expectations and perceptions survey.

1 as 4 and statement 2 as 2, then the score is −2; expectations greatly exceeded perceptions (reality). This indicates that the individual was expecting more than truly exists. This then could become an alignment problem. Negative values indicate that the individual's reality does not coincide with his or her overall expectations. Averaging the responses will only remove the natural variation. Do examine the frequency of the scores. Aligned teams should score between zero and a positive value. A poorly aligned team with negative scores indicates disappointment with reality compared to their original expectations. Obviously, to determine reasons for any dissonance, interview team members for their personal recollections and opinions.

The process of realignment helps team members adapt to project and program realities. Recommend that all team members recommit to the project outcomes/objectives. One useful suggestion is to place the outcome/objective in a conspicuous space so that team members can constantly realign themselves to this concept. Use it as a header or footer on all e-mails. When the team is focused, it is aligned and adaptable.

Because innovation is often a "learn-as-you-go" endeavor, adaptive individuals make the best team members. Adaptive individuals are ones who can easily adjust to changing realities. Often, what seems reasonable one day may seem ludicrous the next day. Changing priorities are to be expected.

Acceptance of Change

When change takes place, people need to adjust, align, and adapt (see Figure 9.14).

- Change can be painful or pleasing, depending on the situation and the people involved.
- Acceptance of change is a measure of acquiescence—accepting the change as something permanent is accomplished when
  - Minimizing obstacles and barriers

Statement Number	Instructions: Check the Box that Best Matches Your Agreement (Disagreement) with Each Statement	Strongly Disagree	Disagree	Neither Disagree nor Agree	Agree	Strongly Agree
1	When change occurs, I am one of the first to embrace it.					
2	When my company announces a change, I believe it will be positive.					
3	The outcomes of change are generally positive.					
4	My organization does not create barriers to change.					
5	People who embrace change quickly are better adjusted.					
6	Change can be positive when barriers are reduced.					
7	I would accept change if there were additional opportunities.					
8	Accepting change can be made easier if management communicates.					
9	If change reduces stress and anxiety, I would accept it.					
10	There is never a feeling of loss associated with change.					

Figure 9.14 Acceptance of change survey. (Developed by McLaughlin, but not yet published.)

- Developing opportunities for growth
- Developing a new "role" for employees
- Minimizing fear, anxiety, and a feeling of loss

To score the survey instrument, assign a numerical value as follows: 1, strongly disagree; 2, disagree; 3, neither disagree nor agree; 4, agree; 5, strongly agree. To obtain a change acceptance score:

- Calculate an average of the 10 responses; expect averages to be close to 3.0.
- Individuals whose average score is less than 3.0 can be expected to have problems adjusting to change.
- Individuals whose average score is greater than 4.0 can be expected to accept change.

Examine the range for each individual's overall score. High variation indicates poor agreement between individual responses. This suggests that change has been an upheaval for individuals.

Those resistant to change will often find innovation to be frustrating. Yet, innovation requires an open mind, dedication to empirical evidence, and an ability to maneuver around obstacles and barriers. Most employees or associates have this ability if guided and trained to deal with ambiguous situations. The ability to deal with and prosper from these situations is a key trait for adaptation.

Step 6: Tabulate and Track Performance

A critical aspect of any innovation project is whether it can deliver a decent payback over time. Organizations invest in innovation for a specific reason (e.g., cost efficiency, value, or effectiveness). Every project must yield a benefit so that its value is established. For profit-driven businesses, it is ROI (return on investment); that is, what is invested is a fraction of the benefit obtained. For the federal sector, the established driver may be harder to define. Yet, something is driving the project, some benefit received by supporting and implementing the project. At this stage, the project must provide a benefit to evaluate its overall efficacy. Any estimate is purely a guess and the accuracy of the guess dependent on the history and experiences with the process. Figure 9.15 describes the process for tabulating and tracking performance.

Because the project is still in the formative stages, it is critical to establish the performance measures (recall the

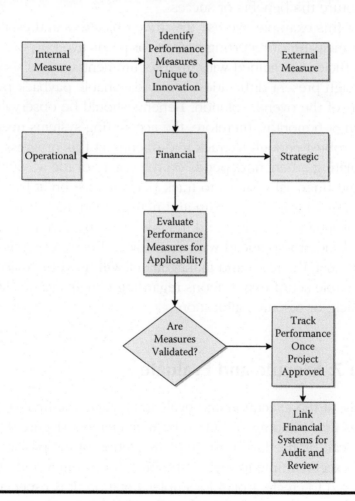

Figure 9.15 Step 6: tabulate and track performance.

discussion of success factors) that define its identity. At this stage, we estimate the benefits and those financial, operational, and strategic measures associated with the project. For this hypothetical case study, performance evaluation includes measuring the benefit, projected costs, and unique measures that define the project.

The team, with management support, reviews the performance measures and establishes the goals that constitute success. Additional work may be needed to clarify these measures and to establish operating guidelines and appropriate systems to capture the benefits of success.

For this example, we use the three objectives and establish a measure of recurring benefit for predictive purposes. The efficiencies gained with the procurement systems should outweigh present difficulties. Some reasonable payback period is part of the overall solution. Benefits should be observable within 3–6 months; therefore, the accounting systems need to be robust enough to track and document this progress. Calculating a benefit depends on the ability of the accounting and financial systems to track performance on at least a quarterly basis. If the team and management feel that the improvement is possible and the outcome/objective attainable, then decide to go ahead with the project. Before executing the project, the team and management will need to devise a reasonable set of expectations regarding benefits value (ROI), payback, or resource allocation.

Step 7: Execute and Evaluate

Finally, at the execution and evaluation stage, the final decision is made and agreed upon by management (Figure 9.16). Plans can be used to begin implementation of the project. This is an excellent time to review the project outcomes and objectives and to create a plan for implementation. It is especially important to review and solidify those project measures used

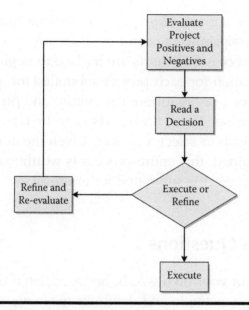

Figure 9.16 Step 7: evaluate and execute.

to identify benefits in step 6 (tabulate and track performance). It is more than likely that a new team is required to move the project forward. Decide what type of innovation works best. For this example, an improvement is warranted. The next chapter details how a project such as this is implemented for success. The key to implementation for innovation efforts is correctly identifying the type of innovation (new, improved, change) required. The process uses a modified version of N²OVATE™ especially designed for the type of innovation and the client organization.

Summary

This chapter introduced an operational process to select an innovation project. Generally, the process takes 1–2 days to complete. A tremendous amount of information is generated concerning the project and is useful for the next (implementation) phase. This chapter was an overview of the process; those who want to investigate the methodology in more detail can contact the authors (drgregmclaughlin@ipsinnovate.com; drbillkennedy@ipsinnovate.com). As with any innovation project, the process flowcharts are a place to begin. We expect some modification for each project submitted for acceptance based on agency or organizational culture and propensity toward innovation. These flowcharts give the reader a detailed process by which to select a project. Given the amount of knowledge gained, this entire process is worth executing each time a macroproject is submitted for approval.

Discussion Questions

1. Discuss, in your own words, an operational definition. Choose an example and define its meaning.
2. Discuss three to five needs that your organization or business faces today. Is there a potential for any of these needs to be an innovation project?
3. Discuss the concept of risk as you understand it. How does risk become a factor in evaluating (selecting) an innovation project?
4. Discuss when, where, and how management will interface with the N²OVATE™ process.
5. Discuss the importance of individual contributions to an innovation project. Which steps rely heavily on human contribution?

Assignments

1. Discuss the process of validating a set of needs (requirements). Remember that at this point, requirements are product specific, based on a contract or agreement. Use an example to proceed through the process.

2. Choose an objective (be sure it is measurable), apply the SMART criteria to the objective, and discuss the results. How much did the objective change?
3. List those financial measures that a project must address to be selected for implementation.
4. Discuss how the N²OVATE™ process could be used by an individual to select an innovation project. What steps would be eliminated, and what others would be added?

Chapter 10

Leading Innovation Project Success: From Concept to Reality

Introduction

On reaching a decision to execute an innovation project, a new process is required to achieve a successful conclusion and make the project ready for implementation. The process offered in this chapter begins where the previous chapter's process ended and is based on innovation type and the N²OVATE™ methodology. As the project begins the implementation phase, the innovation project lead and team can expect modifications to occur in the N²OVATE™ process. These modifications are a natural dynamic in the agility and effectiveness of the N²OVATE™ process and are driven by (1) the fact that the improvement normally varies from normal production efforts and (2) the emphasis on adjustment and adaptation.

Traditionally, the first step in the action phase of implementing a project should always begin with identifying the impact of the planned innovation. To initiate this step, the innovation project team should also identify and label large innovations, which may radically change day-to-day operations, user behaviors, or the legal, social, or econometric environment. We refer to large innovations as macroinnovations and projects with smaller scopes as microinnovations. It is noteworthy that the resulting impact of the project does affect the process used for implementation.

The Impact of Innovation

When most people consider innovation, they think about a "new" discovery, technology, or product (and in some cases, a service). Based on personal experience, people tend to think of products that revolutionized their lives or the environment when considering how to define an innovation. An important consideration regarding macro- (large-scale) innovations is that they can significantly alter the very culture and organization that created or championed a macroinnovation while achieving the goal of satisfying a perceived user (society) need (requirement). These innovations tend to receive the greatest amount of attention and operate at a strategic level in any business, agency, or organization.

More often, macroinnovations can be disruptive, changing existing paradigms, causing markets to realign, and causing organizations to cease or expand by adding to a new set of realities. While the innovation project teams will facilitate these projects, the agency's or organization's management (decision makers) is the key to investing hard-earned cash to make these innovations a reality. This can often result in a tremendous investment of resources (monetary, human capital, etc.) and time to achieve desired goals and objectives. That said, we maintain the position that the majority of innovation projects within the federal government are microinnovations typically affecting not only the user (society/customer) but also the agency or organization at the operational and tactical level.

Research confirms that innovation perceptions differ across organizational cultures, age groups, and ethnicities. These perceptions directly influence consumer behaviors (including use) and purchasing decisions. As a case in point, not everyone wants or needs that latest gadget or technology. Generational (age) differences also influence how a person perceives innovation: Those with more experience will understand innovation as something more than new or novel. These individuals will recognize innovation when something improves or changes for the better (refer to Chapter 2 for further clarification).

If innovation is more dynamic than just a new invention or discovery, then more people can involve themselves in activities that produce innovative outcomes. In reality and from our perspective, innovation can occur on a daily basis. Individuals (employees/associates) direct these projects and are committed to the same standards of excellence. These micro- (smaller-scope) innovations do not disrupt the organization, but they can certainly have game-changing potential for life and reality in general. In micro- or smaller-scope innovation projects, individuals or small teams generally direct (microinnovations). These innovations share many of the same qualities and requirements as a macroproject. Both macro- and microinnovation projects require the following:

1. A process or method to evaluate the project's efficacy (usefulness)
2. Frequent assessment
3. A clear goal or objective, including a defined/accepted need or requirement
4. A relevant benefit (identified success factors)
5. Agreement and support from management (decision makers)

Microinnovations should involve the largest number of employees, suppliers, customers, and stakeholders. These innovations tend to be more relevant to the individual given that the same individuals will see the greatest benefit. Employees will feel empowered, responsible, and engaged. Many microinnovations have led to significant long-term benefits. The 3M Company is famous for its number of microinnovations that have yielded large benefits. Microinnovation is, in fact, "incremental improvement." Incremental innovation fulfills a need by improving the performance of an item. Items could easily be policies, procedures, a process, products, or services. Performance improvements, beyond those attributed to continuous improvement, exceed expectations and thereby qualify as innovative.

Microinnovation can just as easily occur with upstream or downstream suppliers or customers on a specific item-by-item basis. Directing innovative activities that add value toward users improves their perceptions and overall evaluation. This can solidify long-term working relationships and provide the agency or organization with a distinctive competitive advantage. Customers (users) trust those who look after their best interests. Focusing innovations at a customer level could provide a significant competitive advantage, especially if the user experiences these innovations frequently. Apple's strategy for releasing sustained and relatively frequent incremental capability improvements (versions of its iPhone series) have led to strong brand recognition and market and a devoted customer base. It has also spawned a growing market segment of small software and applications development startup opportunities. Another example is the adoption of Microsoft's Windows NT operating system that displaced Novell, Banyan Vines, and Pines in the late 1990s as the government's main networking operating system.

Macroinnovation

Macroinnovations receive the bulk of attention in the media due to the significant benefits and marketplace changes. New discoveries and new technology dominate the headlines. The latest weapon, technological breakthrough, and newly created material tend to challenge our thinking while often meeting or setting the stage for a new set of needs (requirements). Often, the federal sector leads in this type of innovation. The government has financial capability (deep pockets) to support macroinnovation investments. There are numerous implications to national security or threats to a global leadership position if government does not stay on the cutting edge of technology. For example, the rise of sophisticated hackers who can easily tap into highly sensitive materials and knowledge resulted in measures to counteract these intrusions. Security protocols, new technology, and new software were macroinnovations that eventually migrated to the civilian marketplace. Macroinnovations affect a large number of individuals and change the "rules" forever. Numerous methodologies and strategies exist for macroinnovations. One of the more famous strategies is blue ocean, which is characterized and defined by "untapped market space, demand creation, and the opportunity for highly profitable growth" (Kim & Mauborgne, 2005, p. 106). Blue ocean strategies are rare as they change the existing paradigm by creating a new product, technology, or service that never existed previously. Another well-known "macrostrategy" is disruptive innovation, which describes the upheaval that occurs when a "new" product or technology (a discovery) is introduced to the marketplace. The process for disruptive innovations leapfrogs traditional research and development (R&D) efforts. It also resets the existing paradigm but tends to disrupt operations.

Macroinnovations can take many paths and produce many results. These occur infrequently (maybe only once in an organization's history) and are difficult to plan. The scale of macroinnovations, for the federal sector, can occur anywhere along the chain of command. The impact of the projects affects employees, suppliers, and customers (users). These are in contrast with microinnovations, which occur with greater frequency.

Microinnovation

In contrast, microinnovations often begin with objectives that management provides for ongoing improvement (e.g., an agency's or department's continuous process improvement [CPI] programs). Projects can emanate from existing research facilities, engineering, operations, departmental functions, customers (users), an individual. Rather than trying to disrupt or reinvent the marketplace (as macrostrategies do), microinnovation chooses objectives more grounded in the expertise of the organization. Microinnovations can be either large or small leaps forward for the organization. Microinnovations involve the suppliers, customers (users), and stakeholders. Rather than disruptive, it is a "disciplined" innovation approach that follows a set of methodologies to accomplish one or more objectives.

The methodology presented in this chapter focuses on microinnovations with the goal of incremental improvement. This methodology takes many forms and addresses many objectives. For the federal sector, microinnovations would greatly benefit the organizations. The strategy is to bring this thinking, the N²OVATE™ processes, and tools into all departments, no matter what its purpose, goal, or objective—in essence bringing innovation to a more personal level. Unlike the traditional approach that requires the submission of ideas chosen by management, microinnovation can occur more spontaneously, using a disciplined approach to achieve a desired outcome when needed throughout the organization. Individuals can use the process described in the previous chapter to determine the "worthiness" of the project before critical assets and resources are obligated or expended.

Implementing a Successful Project

Once the project receives approval, it is time to implement it. As with any innovation project, conditions and situations may change prior to implementation. Therefore, there are always periodic or recurring negotiations and evaluations that must occur to take the project from concept to reality. To understand the implementation phase, we use the example from the previous chapter. We present two unique scenarios to describe how to implement the case study.

Scenario 1: Incremental Improvement

Considering the case study example in Chapter 9, if the project was approved, the first choice is for an improvement. That is, updating the software would constitute an incremental improvement. The critical element in improvement is performance. If the performance is substandard, then the purpose of the innovation is to raise performance to those levels that exceed expectations. On the other hand, if performance meets expectations but would add more value if it increased significantly, this would also constitute incremental improvement. For the case study, though, the performance is lacking at present. Modifications to the N²OVATE™ process involve the following steps:

1. N = Nominate and negotiate
2. R = Reasons and requirements (needs)
3. O = Operationalize the outcome
4. V = Verify and validate
5. A = Adapt and align
6. T = Track and tie to performance
7. E = Establish controls

In the first step (Figure 10.1), begin with creating a new team. Assess the outcome and determine the resources needed and the benefit expectations. If the project continues to offer the benefits identified in phase 1 (step 6), then continue to pursue a successful outcome. Step 1 can begin prior to implementation. It is also time to reconsider team members,

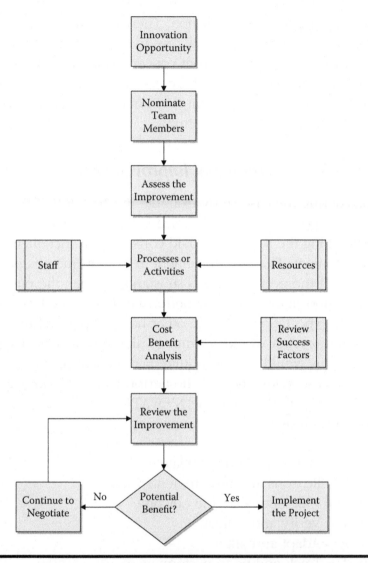

Figure 10.1 Step 1: nominate and negotiate (implementation).

especially those experienced in project implementation. In addition, negotiations continue with management to refine objectives, requirements, and success factors. This stage provides for a review of the benefits provided and the overall contribution made to the agency or organization.

Step 2 (Figure 10.2) involves determining reasons and causes for the lack of performance, allocating resources, and

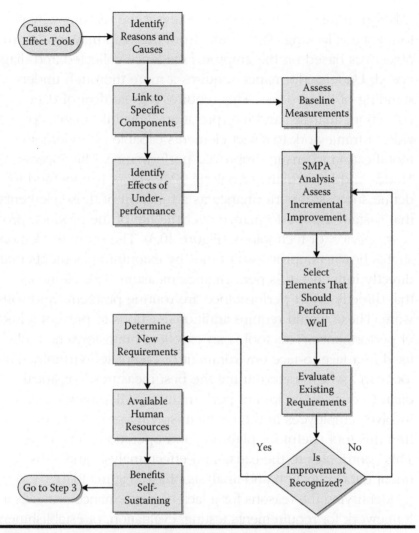

Figure 10.2 Step 2: reasons and requirements. (SMPA = Success Modes and Performance Analysis.)

determining overall requirements. Numerous tools can be used (described in *ENOVALE™: How to Unlock Sustained Innovation Project Success*) (McLaughlin & Caraballo, 2013b).*
This is a lengthy stage, given all the components that must come together to understand the causes and reasons for underperformance. Even though the team identified a possible

---

* A revision of this text is due for publication by CRC Press by the end of 2015.

solution in phase 1, it is crucial to revisit the reasons why performance is lacking. The results from this step may modify the objectives based on the empirical evidence collected and diagnosed. Underperformance requires a more thorough understanding of the process. This requires a great deal of data collection, analysis, and interpretation. This information provides a framework to select elements capable of undergoing modification to ensure improved performance. The Success Modes and Performance Analysis (SMPA) tool is a method to define successful performance as a function of those elements that sustain that performance over the life of the product, process, service, or technology (Figure 10.3). The example demonstrates how to improve wait times by examining elements that directly influence this performance measure. Two elements that directly affect performance are routing problems and software. These would require additional actions to prevent a loss of performance. This tool is an excellent brainstorming tool used in a face-to-face environment or completed virtually. This tool can assist in determining the best measures (empirical data) to control the loss of performance. SMPA analysis also involves employees in the problem-solving process. Clients find this tool useful for planning and operational purposes. This contributes to the cause-and-effect analysis and subsequent data collection and analysis phase (Figure 10.4).

Identifying the reasons for a lack of performance provides a framework for requirements testing, evaluation, or establishment of new requirements. With new requirements comes a new round of data collection, baseline measures, and new measurement systems. The idea is to continue the iteration process until performance has exceeded expectations and is operating at a validation and verification level (new level of sustained performance). As with all ENOVALE™-based methodology, the next step is verification with empirical evidence. Data collection need not be overburdening—collect only "enough" data to detect a change in the process. Be aware that calculations, such as averages, hide information and are difficult to achieve. Watch and

Process or Product Name:									Prepared by:			Page __ of __
Team:									SMPA Date (Orig)		(Rev)	
			Improve Wait Time									
Process Step	Key Process Input	Potential Success Modes	Potential Performance	PRB	Potential Causes	NEP	Current Actions or Controls	SUS	SPN	Actions Recommended		
What is the process step?	What is the component, part, or element?	In what ways can the component, part, or element improve?	What is the effect on performance?	Impact Probability- Rate the chance of continued improvement.	What could cause the component, part, or element to affect performance negatively?	How frequently would a negative affect occur?	What actions (Controls) are needed for this improvement to be sustained?	How well can the improvement sustain increased performance?	Success Priority Number	What are the actions required for maintaining improved performance?		
	Customer's Available Time	Less Wait Time	Judged more efficient	10	Problems with Routing	6	Modify software	5	300			
	Customer's Available Time	Less Wait Time	Judged more efficient	10	Problems with Routing	6	Increase operators	1	60			
	Customer's Available Time	Less Wait Time	Judged more efficient	10	Problems with Routing	6	Increase menu options	10	600	Check feasibility of increased menu selection		
	Customer's Available Time	Less Wait Time	Judged more efficient	10	Software	8	Purchase or design software	8	640	Check available hardware for purchase		
	Customer's Available Time	Less Wait Time	Judged more efficient	10	Human interaction	8	Training, follow-up	6	480			

Figure 10.3 SMPA example (wait times). (PRB = probability of improvement; NEP = negative effect probability; SUS = sustainability unit score; SPN = success priority number.)

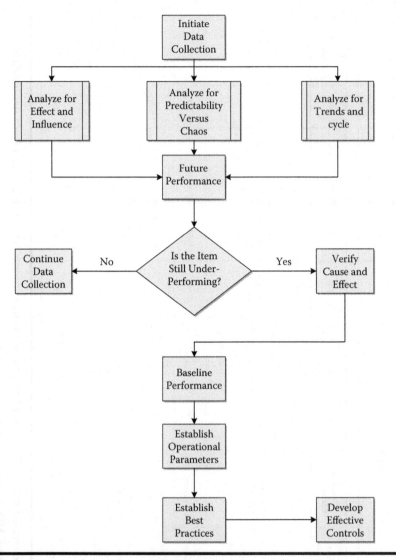

Figure 10.4 Step 3: operationalization.

monitor variation; this statistic describes the consistency (repeatability) of the process. An inconsistent process is one that is impossible to repeat and to forecast.

It is common to shortchange the data collection and analysis steps. In a rush to operationalize the innovation, it is simpler to confirm rather than understand performance. Yet, collecting a small amount of data will verify if the operational parameters are descriptive and functional. Nonfunctional operating limits cause frequent adjustment, slowdowns, and increased losses. Even for processes that are human focused, individuals need to know the parameters (guidelines and requirements) in which they must operate. Estimate and then verify these operational limits (boundaries).

This step can be time consuming (if executed improperly) and difficult to convince management/leadership of its importance. Minimizing this step risks failure because vague operating (requirements) limits result in frequent adjustment (attention), repair, and high maintenance costs. Determining operating limits (and specifications) is both science and art. To summarize, consider these elements as necessary:

1. Processes normally operate within a range; so do people.
2. The range needs to define process operating limits while remaining within the requirements (limits and specifications) nearly 100% of the time.
3. Specification limits are customer or user driven.
4. Limits are set based on empirical evidence.
5. A process can run (operate) outside limits (for a short tine) but with increased risk of failure.
6. There should be a one-to-one relationship between operating limits and user or specification limits; operating limits must ensure that the product, service, or technology meets specifications 99.7% or better.

Services are no different. Generally, operating limits are not part of the conversation when developing services. However, when applied, these ensure nearly flawless execution. The provided flowcharts work well for product (technology) or service. For services, innovation requires a mindset change as well as recognition of a service as a process. In the past, theorists referred to this mindset as systems thinking (today, critical thinking). The three basic elements of critical thinking are "function, structure, and process" (Ing, 2013, p. 528). In common terms, systems thinking is "the outcome, the inputs and components, and the sequence of activities" (Ing, 2013, p. 528). Services contain all three of these elements and therefore can be designed (and developed) to exceed expected performance. Those who apply these criteria will have developed more efficient and effective services while providing an environment for incremental innovation. For the federal sector, this is natural because services are the majority of user offerings.

The next step is one of verification and validation (a central theme). Figure 10.5 details the process of verification. Validation is verification of key operational parameters. It is also time to develop policies and procedures from the validated operating limits. New or modified requirements are instituted prior to rollout (implementation). Operationalization is a step that prepares the product, service, or technology for implementation. If this step is skipped or diluted, the risk of failure increases due to unplanned or untested conditions. The less you know about a process, the greater the chance of errors, mistakes, and miscalculations will be. Prevention is always the best cure.

The next step (Figure 10.6) is one that is often neglected. In this step, the attempt is to focus attention on adapting (aligning) employees to the improvements. It is time not only to celebrate but also to prepare those responsible for their new duties. Here, consensus is key and leadership's role critical. Communication is vital, especially for those with new or additional responsibilities. It may also be time to communicate with suppliers and users (customers), especially if affected by the improvement.

In addition, aligning the employees to the new realities is crucial for success. As innovations become more commonplace, it will become easier to adapt to new outcomes (Figure 10.6). Alignment and finalization of benefit and cost savings (reduction) will enable speedy implementation. With any innovation, there is a comfort factor (zone) that needs to be reached with employees, customers (users), and suppliers.

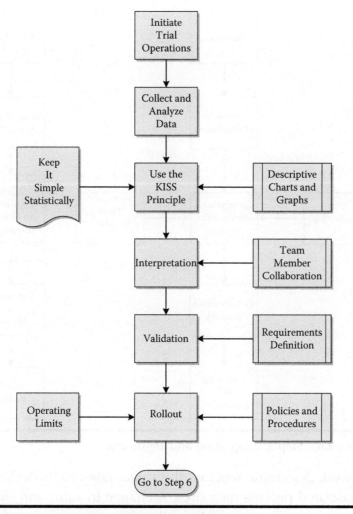

Figure 10.5 Step 4: validation and verification.

Include stakeholders for businesses. Reaching or achieving this comfort zone achieves a level of acceptance, which is critical for sustaining success.

As operations come "online," outcome performance (financial, operational, strategic) measures must begin functioning as well. The process (Figure 10.7) begins by reviewing these measures for both relevancy and use. At this step, improvement in overall value is the key. If efficiencies, effectiveness, or cost savings improve, there must be an associated value. Once

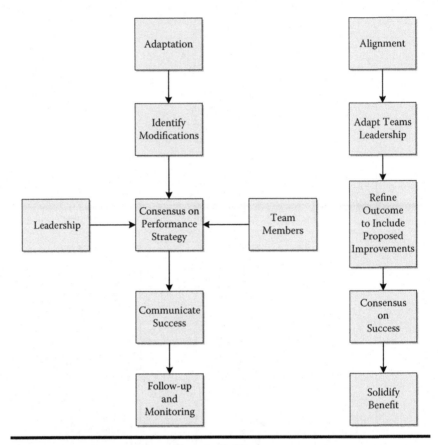

Figure 10.6　Step 5: adaptation and alignment.

reviewed, determine whether the value measure links directly to associated process measures. A change in value can signal a change in performance. As well, subtle changes in performance can easily indicate movement in value.

Next, fully align these measures to financial and accounting parameters associated with project management (Figure 10.7). Use this step as a validation of these measures and indicators. Determine what to track and tie it to those established measures of performance. Finally, the project becomes reality. Figure 10.8 details the steps in initializing the project as operational.

At this point, the discussion and process flow is complete. As with every process, there are far too many unknowns to have a "one size fits all." Each process is unique, yet fundamental issues remain constant. This process deals with these issues that leaders and executives must manage. Failure to do so increases risk and failure.

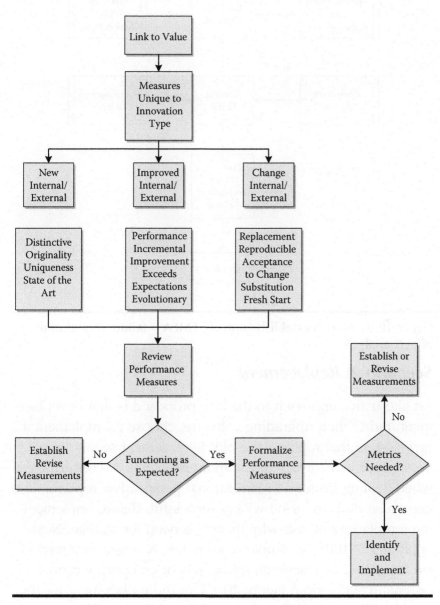

Figure 10.7 Step 6: track and tie to performance.

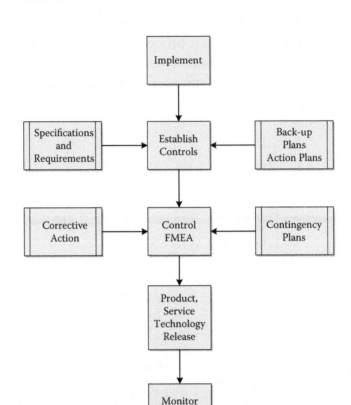

Figure 10.8 Step 7: establish controls. FMEA = failure modes and effects analysis.

Scenario 2: Replacement

An alternative approach to the one proposed is that of replacement; rather than upgrading software, choose to implement a new system that replaces the old. Replacement seems simple. We replace items every day without problems or concerns, yet when dealing from an organizational perspective, replacement can cause disarray or be widely successful. Before implementing a replacement, ask why there is a need for change. Next, apply the SNIFF (S = simple or complex; N = new or recycled; I = integrated or reactionary; F = facts or emotion; F = final or tentative) test from Figure 10.9. Can the replacement meet these criteria and add value?

Leading Innovation Project Success: From Concept to Reality ■ 215

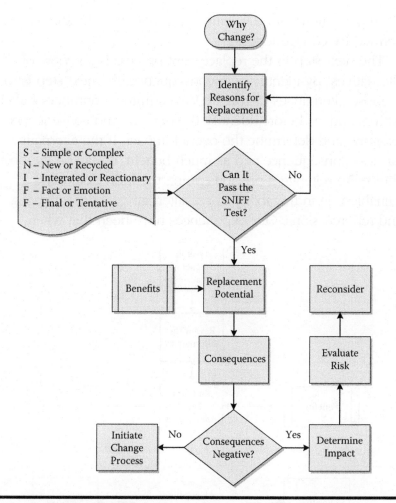

Figure 10.9 Step 1: replacement cycle.

Next, consider the consequences of replacement: Will a new procurement system meet all the existing needs? Describe the additional benefits and consequences associated with the replacement. Negative consequences carry the greatest risk. A new procurement system would solve some problems, but consider the issues of compatibility, training, and experience. If the new system is not fully compatible, then this is a negative consequence. The risk could be such that the system remains nonfunctional until compatibility issues are resolved. Consider consequences of any replacement. Step 1 evaluates the replacement and determines the benefit/risk ratio as defined by consequences.

The next step in the replacement process is a review of alternatives. By identifying a consequence, the next step is to propose alternatives (Figure 10.10). Examine a number of alternatives and, as before, identify the benefits and consequences—examine and determine the overall impact. If the alternative has less consequence and as much benefit, it may be the best choice. To reiterate a previous point, the individual plays a valuable role in the ability to see alternatives, consequences, and repercussions. Our experiences have been that when

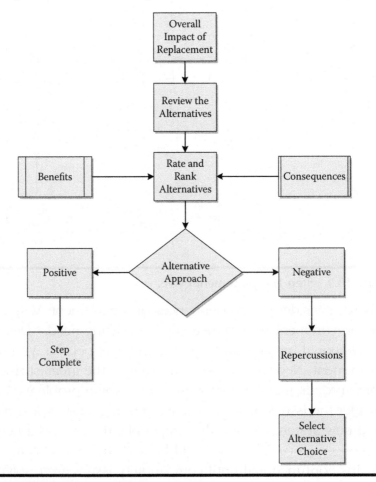

Figure 10.10 Step 2: replacement cycle.

management wants to have innovation and has established a particular benefit, there is often no consideration of alternatives.

Consider an example. A large organization wanted to outsource its procurement department. They conducted research and determined that an Asian country would be the most cost-effective location to conduct procurement affairs. The benefit drove the implementation. No one ever considered the risk involved in moving an entire department to a foreign country or if a better alternative existed. The organization moved the department to Asia only to find that the new personnel received inadequate training and lacked the overall experience to run a procurement process. In fact, rather than saving money, the cost of the department doubled. No one considered the consequences that can occur when moving a complete department from one country/culture to another. Finally, the organization had to reconsider and moved the procurement process back to its original location. This is a perfect example of both consequences and repercussions when failing to identify alternatives.

Step 3 (Figure 10.11) reinforces the need to consider and plan for possible repercussions. Using the AREA (alternative repercussions evaluation analysis) template is a key. *ENOVALE™: How to Unlock Sustained Innovation Project Success* (McLaughlin & Caraballo, 2013b) contains the information for constructing such a diagram. Figure 10.12 is an example of an AREA spreadsheet. Allot 1 hour to complete this critical activity. The AREA template enables the team to brainstorm ideas concerning alternatives, consequences, and repercussions. The AREA diagram helps by identifying alternatives to a solution and trying to determine what would make it fail (a repercussion). This helps to establish controls and evaluate concepts previously undiscovered. It also teaches this type of thinking: Always consider there is never a "right" answer, only choices and alternatives. Remember, consequences and repercussions are associated with all processes, especially those designated for change.

218 ■ *Innovation Processes and Solutions in Government*

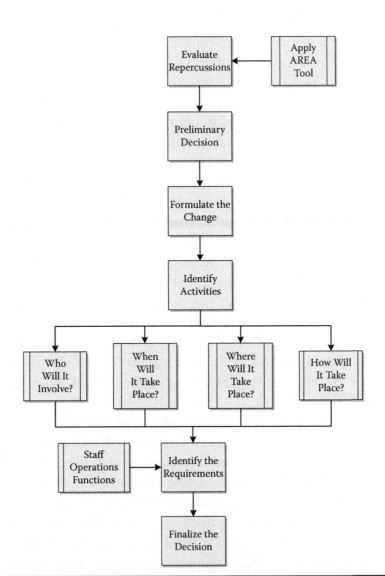

Figure 10.11 Step 3: replacement cycle. (AREA = Alternative and Repercussions Effects Analysis.)

The purpose of this tool is to highlight alternatives, consequences, and possible repercussions. You begin the process by choosing the consequence of a particular action. Next, identify possible alternatives and repercussions. Identify the impact (influence) on the consequence. Then, look for reasons and root causes. Finally, determine what controls are in place to

Process or Product Name:											Prepared by:				Page : _____ of _____	
Responsible:											FMEA Date (Orig) : _____ (Rev) _____					

Alternative and Repercussion Effects Analysis
(AREA) Note: Failure = Effect of the Repercussion

Process Step	Alternative	Repercussion	Potential Failure Effects	S E V	Potential Causes	O C C	Current Controls	D E T	R P N	Actions Recommended	R E O C	Resp.	Actions Taken	S E V	O C C	D E T	R P N
Identify the process step or outcome, or consequence.	What is the alternative?	Identify the repercussions (failure points).	What is the effect on the consequence?	How severe is the effect to the project outcome?	What are the causes or reasons for this effect?	How often does a cause occur?	What are the existing controls and procedures that minimize the effect on the outcome?	How well can you detect the cause?		What are the actions for reducing the occurrence of the cause or improving detection?		Who is responsible for the recommended action?	What are the actions taken with the recalculated RPN? **Be sure to include completion month/year**				
									0								0
									0								0
									0								0
									0								0
									0								0
									0								0

Figure 10.12 AREA spreadsheet. (SEV = severity; OCC = occurrence; DET = detectability; RPN = risk priority number.)

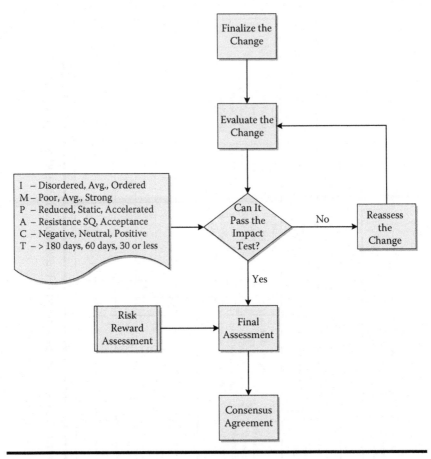

Figure 10.13 Step 4: replacement cycle.

prevent the repercussions from influencing the consequence (outcome). Understanding what affects the consequences will help in choosing the best (or no) alternative.

Next, verify the change and determine its overall impact. In this step (Figure 10.13) the replacement (change) is reevaluated using the IMPACT (Integration, Managed change, Performance, Acceptance, Communications, Time) test:

I = disordered, average, ordered
M = poor, average, strong
P = reduced, static, accelerated

A = resistance, status quo (SQ), acceptance
C = negative, neutral, positive
T = more than 180 days, 60 days, 30 or fewer days

This is another test that is not empirically based but coordinates the knowledge and experiences of those most involved with the replacement. It is also another opportunity for a risk/reward scenario analysis. Risks and rewards are another topic for a future book. Risk involves any incidence that could result in failure. The team can verify the existing replacement or its alternative as acceptable and validated. Trying to implement a replacement and then fixing it later is an expensive proposition.

A difficult part of replacing a familiar process is the change in outcomes, process, and personnel. During these times, there is a need for alignment, acceptance, and acquiescence. The alignment stage helps employees, suppliers, and customers accept and use the new process. Aligning to the decision is critical and must precede the acceptance phase (Figure 10.14). This will mean developing a strategy for handling resisters. Keys are convincing them through success, giving them some time to adjust, and letting them be part of the implementation.

First, align the decision to all external stakeholders. Demonstrate the benefits, discuss the advantages, and acknowledge any disadvantages. Second, align internal personnel using their experience and knowledge to implement a solution. Finally, allow time to align and accept the decision. Take action to minimize resister effects. This can include moving the individual, further training and assistance, or removal.

Leadership (Figure 10.15) is critical for replacement success. Leaders provide a vital role and channel that includes the open, two-way communications required for success. Leaders provide the road map for success. Employees will

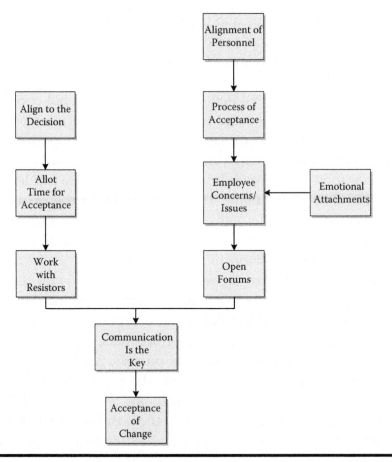

Figure 10.14 Step 5: replacement cycle.

follow leadership when the benefit is clear and objective accomplishable.

Leading an effort inspires those who support the effort. Leadership provides stability, vision, and purpose. Employees who see (and experience) strong support will trust the leader's decision.

Finally, embrace the change, and the change will become permanent (Figure 10.16). This is the rollout phase. Accept feedback from those the process "touches" as they have a unique insight. Ask experienced people to validate the decision—let them provide the "best" reasons for the replacement.

Figure 10.15 Step 6: replacement cycle.

In addition, let employees adapt to the replacement, accept the decision, and share in the benefit.

Summary

Organizations need both macro- and microinnovations. These two innovation types are the start of a well-balanced innovation strategy. Repeated successes with microinnovations set the stage for the more elusive

Figure 10.16   Step 7: replacement cycle.

macroinnovation events, which occur much less frequently within a federal agency's or organization's history. Both micro- and macroinnovation successes address a number of human resources issues related to employee productivity, creativity, and self-esteem. If individuals know they can innovate, they will; this will ultimately help the organization improve, grow, and improve competitive value.

After selecting a project, management and the team can move forward with implementation. This is the stage when the item goes from design to full development. Depending on the type of innovation, the resources required, the difficulty, and the impact on the organization, implementation can extend over a period of weeks to years. This time component may cause the team to develop an alternative approach. Alternatives have consequences and repercussions. Considering an alternative permits the opportunity to explore options and new methods and refine objectives. After finalizing the decision, management must lead the effort, define the benefits, and consider the fate of employees impacted by the decision. These actions will influence the organization well into the future. Work to align stakeholders (both external and internal) and provide open communications and feedback. Give employees time to adjust and accept the decision and recognize and highlight the benefits.

Discussion Questions

1. Give an example of a macro- and a microinnovation. What is the most notable difference (in your opinion) between these types of innovation?
2. Discuss, in detail (i.e., explain each step), the reasons and requirements flowchart (Figure 10.2). What does this step accomplish?
3. Why is data important in making a good decision? Why is data critical in monitoring and controlling various project success factors?
4. Why is an alternative often preferable to an existing outcome? Why is it critical to understand the role of consequences and repercussions?
5. Discuss why leadership is critical after a decision is made but prior to full implementation?

Assignments

1. Propose a macroinnovation for your organization (assume it has passed the project selection phase). Detail the resources, staff, and customer (user) requirements needed to begin implementation.
2. Consider a microinnovation from your perspective. What, if improved, would address an unsatisfied need and deliver a benefit to the organization. How would you plan to implement such a project?
3. Why consider consequences when implementing a project outcome? Consider a macroinnovation such as the Apple

iPad (or any suitable alternative). What is a viable consequence of using the innovation (may be positive or negative)? How would this generate the need for an alternative?
4. Consider the replacement strategy detailed at the end of this chapter. Can you identify an application for this in your organization? If so, compare the flowcharts to what actually occurred in your organization. Which process would produce the best results?

Chapter 11

The Federal Government: 2015 and Beyond

Introduction

As innovation can be a fleeting opportunity based on timing, dedication, and support from decision makers, the future of innovation in the federal community will remain prominent. Many agencies use *innovation* to prescribe their very activities and strategy to secure future funding in support of achieving their core competencies; others use the term in a cavalier manner to describe the indelible marks they have sought to achieve in their own individual careers. The central questions remain:

1. Can you be innovative in your present position?
2. From your perspective, is your department or agency innovative?
3. What are the top three innovative initiatives you or your agency completed over the past year?
4. What value did those innovative initiatives or projects bring to the agency or organization?

We understand that immediate and dramatic course changes will not be easy, and they will take time. The impacts will be far-reaching and will not be popular as change naturally conjures uneasy feelings for almost everyone. It will take a paladin with the right wisdom, vision, courage, and commitment to navigate those dramatic changes in migrating the nation toward a smaller government operating on a fraction of the current budget. Beyond the innovation evangelist, the cuts in funding will drive innovative and creative measures for keeping the government stable as the workforce adjusts in size and to a leaner mentality with a more youthful presence. There is no panacea or easy path down this road, and there will be a lot of angst and pressure that will test the very mettle of long-time government professionals who have served this great country. The government culture must change and become more agile and flexible; those who have already sacrificed much in the service of this country must be willing to sacrifice more to those in this nation.

Unlike the military force structure that went through many buildups and drawdowns with each conflict starting from the Korean War, the civilian workforce only grew in size. The days of reckoning are here, and a gradual drawdown can likely be achieved by the 76 million baby boomers and veterans in today's workforce deciding it is time to leave the workforce and retire. That said, over 70 percent of those boomers and veterans will be hard pressed to leave the job market based on the reality of retiring (current estimates provide there are only some 47 million millennials, generation X, and generation Y resources to fill those gaps). The opportunity exists for the government to reduce the workforce and not replace those workers over the next 5 to 10 years. Combine this with the advances of information technology and communications, the need to travel or work in an traditional office space rather than a home office will test the traditional mindsets of senior government leaders, who have always worked in a brick-and-mortar office surrounded by a large supporting cast. The advent of temporary or seasonal positions versus full-time positions within government has been used for some time (e.g., National Parks, Department of Agriculture, Department of Defense [DoD], etc.).

We expect that by 2025 a reduction of 30 percent of the government workforce will occur. Of those remaining, some 35–40 percent will work from home 2 or 3 days a week. Service functions will remain staffed; however, administrators may eliminate many manual functions that required a physical body in place to perform such functions as checking identification or receiving payments). Those who do remain will have seasonal positions focused on providing college students and retirees an opportunity to gain work experience, offset loans, or bridge the increasing gap between pay and inflation (the cost of living). Our educational institutions must also change their philosophy and focus on the very community that keeps this nation prosperous: the small-business community. Not every successful college graduate will find a position in a Fortune 500 company or even in the discipline studied in college. In essence, the days of retiring with a government pension will be an oddity because most permanent positions will have their retirements funded from a civilian-based source versus the government coffers.

We also foresee a single military academy for training military leadership, with core military curriculum taught in the first two years and service-specific curriculum in the final two years. This will resemble the institutions that have separate undergraduate and graduate divisions. The administration and oversight for command and staff colleges, war colleges, and senior civilian development courses will also be under one administrative function. Moreover, specific military departments could be reassigned under a joint organization and the administrative staffs reduced as the size of the forces decrease or are shifted to the guard or reserve components over the next 10–20 years. Most would agree on not compromising national security; however, there are some things we no longer can afford to commit to if we do not want to burden our children and grandchildren with unmanageable taxation and a life without opportunity.

Our past peacekeeping efforts have cost the taxpayer dearly, and the realities and necessities of intervening in a world where violence, strife, and unconscionable atrocities were justified. However, the tragic loss of the great American heroes who have answered duty's call across our long history as a nation and the realities that we can no longer support every event abroad without maintaining a sizable humanitarian and military capability will always run contrary to what sequestration is driving. This book is not about politics or social and economic reform; it is about innovation, and it is time for the government to act responsibly and with open and transparent purpose.

With the saturation of the cybersphere and the satellite networks that support the neural network, the increasing reliance on real-time connectivity, anywhere and anytime, will continue to challenge the everyday smartphone or tablet user. The Federal Aviation Administration's (FAA's) terrestrial networks for air traffic management will give way to the satellite-based guidance (GPS) systems, for which highways in the sky will function much like those on Earth. These systems can identify the location of an aircraft within a yard of its actual location over land or sea and allow private and public transportation providers to optimize their travel times by sharing tighter airspace corridors. Along with the demand for the government to release more of the frequency spectrum (i.e., wireless broadband) they own for government agency use, expect the Federal Communications Commission (FCC) and FAA to place increasing controls on how available bandwidth is allocated, when it is used, and how it is used. One might argue that it may be innovative for the government to release unused bandwidth for use by commercial providers, which could generate income for the government.

The advent of the unstaffed aerial vehicles (UAVs) using GPS technology has opened the door for pushing heavy transportation aircraft into new dimensions. Beyond the thought of large dirigibles employed to economically carry sizable, non-time-sensitive payloads to disparate areas across the globe, to minimally staffed crews supporting transoceanic flights, these are simply incremental improvements based on the demand for resource preservation and improved efficiencies.

A centralized healthcare system leveraged on Medicare and Social Security will only further trouble the financial challenge the country faces today. We have become a country of emotion and distrust. We do not trust the very politicians charged to care for the welfare of our nation, yet only half of us are willing to vote.

Sequestration has also hit the military and Veterans Affairs (VA) healthcare systems. The drawdown of the military healthcare system and leveraging of specialty care to private providers will continue. However, the carte blanche approach will give way to approaches that are more innovative to manage diminishing resources by pulling many of those specialty care programs and disciplines back to military hospitals. There are an estimated 30 million veterans; almost 10 million of these are eligible for military healthcare at an annual estimated cost of over $50 billion dollars (US Census Bureau, 2010). These veterans are separate from those entitled to use the VA services. The VA is also suffering to find ways to handle the demands of sequestration and the growing populace eligible for its services.

In essence, healthcare accounts for almost half of military personnel spending and almost 10 percent of the annual defense budget. Couple that with the reality that most military retirees, those active, and their families are staying with TRICARE (health plan run by the DoD) instead of private coverage. Many believe the desire to remain with TRICARE is chiefly due to private industry not offering suitable healthcare coverage and plans and those that have plans have raised their premiums, driving the cost for coverage in the private sector to at least 10 times the monthly average of the TRICARE option. Due to increased life spans, expect the DoD to seek innovative approaches in addressing this disparity beyond increasing TRICARE premiums over the next 10 years to at least three times what many veterans and service members are paying today.

Expect some government agencies to reduce their workforce by about 30 percent through attrition and hiring caps over the next 8 to 10 years. The federal government will focus more on their core competencies and become more judicious on what innovative measures and programs will address capability gaps and deliver those capabilities on a shorter timeline. Expect a full and continuing review of their current requirements. New boundaries and definitions are set to decide what programs address burning concerns and capability gaps. Redundant capabilities are required for critical mission sets, but they are not required for every agency or organization. A full and transparent review of assigned programs, responsibilities, mission set, and tasks across the government enterprise will become an ongoing measure to reduce overlapping capabilities across agencies and departments over the next 5 to 10 years. Yearly cuts will result in a decrease of some percentage levied across all government entities that has typically been the approach by the government. That approach simply tells the taxpayer and government entity leadership that we need to prioritize those requirements using factual data.

We simply feel the leadership paradigm across many government agencies and organizations needs to change, as does the role of the leader. The military has long been associated with the statement "an autocratic society supporting democratic principles." Unfortunately, it is our determination that the old mindset of "direction, command, and control" must be less the primary means to address daily operations and more the migration toward one of facilitation, guidance, and support for their primary resource.

We would argue that with a lot less money in the coffers and a significantly reduced workforce on the near horizon, these types of leaders (if they are still around) will find the old-school approach less accepted by a new generation entering the workforce today (the millennials, generations X and Y). The new generation does not operate the way the boomers and veterans do. Regardless, today's government leaders at all levels may want to consider a new strategy for answering some tough questions on how capability gaps are growing and why their agencies or organizations are having challenges meeting their primary responsibilities and tasks. The excuse that we do not have the resources will only work so long before answers are required and leaders are forced to go to their primary resource for innovative thoughts on how to rebuild the government through Lean principles, innovative processes, and perhaps a complete overhaul of the leadership and succession-building process.

It does not take a crystal ball or a study of the military drawdowns since the Korean War to understand this situation. Most agree that the advent of information technology and private-sector participation were among the chief contributors. People with great ideas are the key drivers, and they will continue to be those drivers that will make or break the government's long-overdue transformation. The old days of cardinal rules, unquestionable direction, and scripted processes that often changed will soon disappear or change substantially. In other words, leaders in the hierarchy reformed the ideas based on their own knowledge and experience, causing the potential value or intent to diminish or lose value because of their own paradigm or filters.

In summary, it should not matter who receives the credit for incremental innovation, whether leader, manager, or employee. This is not to say that developing or evolving innovation cannot improve its value; we just recommend accomplishing this in an open and transparent manner through an established process. The government is ripe for a new innovative set of approaches starting at the bottom of the pyramid and where the rubber meets the road: the individual performing the tasks at every level. Your innovation teams should always include members who fit the profiles we have offered in this book.

Summary

Change is the operative word when discussing government today. Change can be dramatic or change can be incremental. Dramatic change can have great benefits or tremendous challenges. Therefore, it is best to subscribe to an approach that gradually, but instinctively, changes to produce ongoing value. One such way of adding value is through the incremental innovation described throughout this book. The process is incremental and does not require change overnight.

For an organization to achieve sustained innovation success, it must evolve from its present state. Incremental innovation is one such approach that provides an evolutionary means to become an innovative organization. We know that the journey must begin, given the constraints that the government faces today. We recommend this evolutionary approach because it delivers sustained success.

References

Aslakson, R. A., Schuster, A. L. R., Miller, J., Weiss, M., Volandes, A. E., and Bridges, J. F. P. (2014). An environmental scan of advance care planning decision aids for patients undergoing major surgery: A study protocol. *The Patient, 7*(2), 207–217.

Atwater, M. M., Lance, J., Woodard, U., and Johnson, N. (2013). Race and ethnicity: Powerful cultural forecasters of science learning and performance. *Theory into Practice, 52*(1), 6–13. doi:10.1080/07351690.2013.743757.

Baregheh, A., Rowley, J., & Sambrook, S. (2009). Towards a multidisciplinary definition of innovation. *Management Decision, 47*(8), 1323–1339.

Berson, Y., Oreg, S., & Dvir, T. (2007). CEO values, organizational culture and firm outcomes, *Journal of Organizational Behavior, 29*(5), 615–633.

Bevanda, V. and Turk, M. (2011). Exploring semantic infrastructure development for open innovation. In *Proceedings of the International Scientific Conference,* ed. M. Mirkovic. Pula, Croatia: Juraj Dobrila University of Pula, Department of Economics and Tourism, 363–386.

Brand, R. F. (2010). *Robert's Rules of Innovation.* Hoboken, NJ: Wiley.

Buisson, B. and Silberzahn, P. (2010). Blue ocean or fast-second innovation? A four-breakthrough model to explain successful market domination. *International Journal of Innovation Management, 14*(3), 359–378.

Capozzi, M. M., Dye, R., and Howe, A. (2011). Sparking creativity in teams: An executive's guide. *McKinsey Quarterly*, April, 74–81.

Caraballo, E. and McLaughlin, G. 2012. Perceptions of innovation: A multi-dimensional construct. *Journal of Business & Economics Research*, *10*(10), 1–16.

Christensen, C. M. (1997). *The Innovator's Dilemma. When New Technologies Cause Great Firms to Fail.* Boston: Harvard Business School Press, 27–32.

Clark, Gordon M. (2008, March 28). Statistical Thinking to Improve Quality. http://www4.asq.org/blogs/statistics/2008/03/. Accessed January 15, 2015.

Clausen, T., Pohjola, M., Sapprasert, K., and Verspagen, B. (2012). Innovation strategies as a source of persistent innovation. *Industrial & Corporate Change*, *21*(3), 553–585.

de Mooij, M. and Hofstede, G. (2010). The Hofstede model. *International Journal of Advertising*, *29*(1), 85–110.

Evans, C. and Wright, W. (2009). The "how to ..." series. *Manager: British Journal of Administrative Management*, (65), 10–11.

Federal Knowledge Management Working Group. (2011). Community of practice. NASA.GOV. Retrieved from https://google.com/site/fmwgroupnasa/community-of-practice.

Ghazinoory, S., Abdi, M. and Azadegan-Mehr, M. (2011). SWOT methodology: A state-of-the-art review for the past, a framework for the future. *Journal of Business Economics & Management*, *12*(1), 24–48. doi:10.3846/16111699.2011.555358.

Hagel, J. and Seely Brown, J. (2005). Productive friction. How difficult business partnerships can accelerate innovation. *Harvard Business Review*, February, 83–91.

Higgins, J. M. (1995). *Innovate or Evaporate: Test and Improve Your Organization's Innovation Quotient.* Winter Park, FL: New Management.

Ing, D. (2013). Rethinking systems thinking: Learning and coevolving with the world. *Systems Research & Behavioral Science*, *30*(5), 527–547. doi:10.1002/sres.2229.

Isaksen, S. G. and Ekvall, G. (2010). Managing for innovation: The two faces of tension in creative climates. *Creativity & Innovation Management*, *19*(2), 73–88.

Kennedy, W. R. (2014). Individual (Personal) Perspectives on Innovation: Federal Knowledge Management Working Group. PhD diss., Capella University. ProQuest, UMI Dissertation Publishing (3611870).

Kim, W. and Mauborgne, R. (2005). Blue ocean strategy: From theory to practice. *California Management Review, 47*(3), 105–121.

Kim, W. C. and Mauborgne, K. (2006). *Blue Ocean Strategy*. Boston: Harvard Business School Press.

Kim, W. C. and Mauborgne. R. (2009). How strategy shapes structure. *Harvard Business Review*, September, 73–80.

Martin, J. (2011). Dynamic managerial capabilities and the multibusiness team: The role of episodic teams in executive leadership groups, *Organization Science, 22,* 118–140. http://orgsci.journal.informs.org/content/22/1/118.full.pdf+html. Accessed December 2, 2011.

Masters, C. (2007). How Boeing got going. *Time, 170*(11), 1–6.

McCreary, L. (2010). Kaiser Permanente's innovation on the front lines. *Harvard Business Review, 88*(9), 92–127.

McGrath, R. G. (2011). Failing by design. *Harvard Business Review, 89*(4), 76–83.

McLaughlin, G. (2012). Why Innovation Is So Often Hit or Miss. Innovation Management.com. http://www.innovationmanagement.se/2012/06/25/why-is-innovation-so-often-hit- ormiss/. Accessed October 31, 2013.

McLaughlin, G. and Caraballo, E. (2013a). *Chance or Choice: Unlocking Innovation Process*. Boca Raton, FL: Productivity Press.

McLaughlin, G. and Caraballo, E. (2013b). *ENOVALE™: How to Unlock Sustained Innovation Project Success*. Boca Raton, FL: Productivity Press.

Northwest Ontario Innovation Centre. (2015). Types of Innovation. January 21. http://www.nwoinnovation.ca/article/types-of-innovation-355.asp.

Phelps, (2010). A longitudinal study of the influence of alliance network structure and composition on firm exploratory innovation. *Academy of Management Journal, 53*(4), 890–913.

Porter, M. (1987). From competitive advantage to corporate strategy. *Harvard Business Review, 65*(3), 43–59.

Poškienė, A. (2006). Organizational culture and innovations. *Engineering Economics, 46*(1), 45–50.

Raynor, M. E. (2011). Disruption theory as a predictor of innovation success/failure. *Strategy & Leadership, 39*(4), 27–30.

Richardson Jr., C. W. (2012). Diversity performance as a factor in marketing programs: A comparative analysis across ethnic group target audiences. *Journal of Marketing Development & Competitiveness*, 6(5), 62–70.

Shook, John (2014). Transforming Transformation. http://www.lean.org/shook/DisplayObject.cfm?o=2533. Accessed December 14, 2014.

Teece, D. J., Pisano, G., and Shuen, A. (1997). Dynamic capabilities and strategic management. *Strategic Management Journal*, 18, 509–533. doi:10.1002/(SICI)1097-0266(199708)18:7<509::AID- SMJ882>3.0.CO;2-Z.

United States Census Bureau (2010). United States census 2010. Retrieved from http://www.census.gov/2010census/data/.

Zhuang, L., Williamson, D., and Carter, M. (1999). Innovate or liquidate—are all organizations convinced? A two-phased study into the innovation process. *Management Decisions*, 37(1), 57–71. doi:10.1108/00251749910252030.

Zollo, M. and Winter, S. (2002). Deliberate learning and the evolution of dynamic capabilities. *Organization Science*, 13(3), 339–351.

Index

A

acceptance, *see* IMPACT test
accountability, 55
A checks, 169
acquisitions, choices, 31
active data sources, 134–135
adapt, *see* Align and adapt
add-ons, 81
ad hoc innovation strategy, 146, 147
affiliate cultures, 49, 51
affiliations, 43, 45
age
    differences, 199
    variables, 22–23
aircraft parts procurement case study, 168
Air Force, 88
Air Force One, 8
Air Force Two, 8
airline examples, 27, 34
Alena company, 145
Alice in Wonderland, 129
align and adapt
    incremental improvement, 210–212
    innovation readiness, 117
    project selection, 185–190

alignment
    innovation readiness, 117, 139
    innovation strategy compatibility, 153–155
    project selection, 185–190
    realignment, 188
alliances, public domain, 55–57
alternative repercussions evaluation analysis (AREA), 217–219
alternatives review, 216
American Taxpayers Relief Act (2012), 91
Analysis of Alternatives (AoA), 87
annual reviews, 160
AoA, *see* Analysis of Alternatives
Apple company
    existing strategies, 144–145
    innovation impact, 200
    long-term alliances, 56
AREA (alternative repercussions evaluation analysis), 217–219
Army, 88
Asian values, 45–46
assembling, innovation strategy, 158–159
assessment, *see* Organizational assessment

240 ■ *Index*

assessment-type data, 135
assignments, *see end of each chapter*
assumptions, 179–182
attitudes, 124–125
attributes, linking to, 134
austerity, 3
availability of resources, 48

B

Baker, Jeremy (Jerry), 97
balance, creating, 53–54
bandwidth allocation, 230
Banyan Vines, 200
beauty example, 17
behaviors
    affiliate culture, 51
    predicting, 36
    routines, 65
beliefs
    filtered knowledge, 33–34
    opinions based on, 77
belonging, 43
benefits
    calculating, 192
    process mapping, 70–72
best case, building, 95–96
best practices, 69–70, *see also* Process mapping
bias, 132
"big data," 32
big picture, 59
birth data variable, 22–23
blue oceans
    existing strategies, 143–144
    focus on customer, 54
    innovation types, 20
    macroinnovations, 201
Boeing company, 8, 68, 145
bone diagram, 75–76
brainstorming, 157
Budget and Accounting Act (1921), 90
Budget Control Act (2011), 91
buildup, military, 228
bureaucratic culture, 46
buy-in, 95, 98, 139

C

capabilities, *see also* Dynamic capabilities
    acquiring, 85–86
    addressing gaps, 85–86
    assessment, 26, 169
    Lean model, 73
capability development document (CDD), 87–88
capability production document (CPD), 88
cardinal rules, 233
case, process mapping, 68–70
causal inquiry, 138
cause and effect
    dynamic capabilities, 74–78
    incremental improvement, 206
CDD, *see* Capability development document
Central Intelligence Agency (CIA), 151
centralized healthcare system, 231
certification, 13
champion, Lean model, 74
Chance or Choice: Unlocking Sustained Innovation Success, 168
change
    acceptance, project selection, 188, 190
    agents, 53
    positive outcome, 20
    priorities, 188
charter, FKMWG, 2
chief knowledge officer (CKO), 2

choices
    influence, innovation evaluation, 16
    perspectives, 30–31
CIA, see Central Intelligence Agency
CKO, see Chief knowledge officer
CLINs, see Contract line item numbers
closed-society approach, 99
coalescence of perspectives
    basic concepts, 15–20
    choices, amalgamation of, 30–31
    experiences, 34–36
    inclusion, 38
    information importance, 31–32
    judgment, 33–37
    knowledge, 33–34
    needs/requirements satisfaction, 25–27
    performance, 36–37
    research study, 20–25
    satisfaction, 25–27
    summary, 38–39
    user/customer satisfaction, 27–30
Coast Guard, 88
collaborative culture and environment, 53
collaborative decision making, 13
collectivism, 43
commitment, competency, 160
communications, see also IMPACT test
    political systems influence, 48
    strategy success elements, 158
    success factor, 67–68, 99
compatibility, outcomes and functions, 153–155
competencies, strategy combination, 160, see also Core competencies
competitiveness (adding value), 35
completion criteria, 164

complexity
    compatibility, 153–154
    management, 90–91
conflict minimization, 55
conforming, pressure to, 51
"connecting the dots," 138
consensus building, 13
consequences, replacement, 215–216
consistency, 208
construction, innovation strategy, 155–159
consumer behaviors, see also Customers
    affiliate culture, 51
    predicting, 36
continuing education, see Dynamic capabilities
continuous process improvement (CPI), 202
contract line item numbers (CLINs), 85
contributor categories, 76–77
"cool factor," 144
copyright infringement, 56
core competencies, see also Competencies
    culture and environment, 58
    needs and requirements satisfaction, 82–85
    understanding and appreciation, 42
corporate social responsibility/performance, 44–45
cost effectiveness (profitability), 35, 175
costs/savings, 162
CPD, see Capability production document
CPI, see Continuous process improvement
critical thinking, 209–210
cross-border integration, 42, 64

culture and environment
  affiliate cultures, 49, 51
  alliances in public domain, 55–57
  basic concepts, 5–6, 41–42
  commonality, 36
  estimating effect of, 52–57
  ethnicities, 44–46
  evaluation, 112–114
  family, 51
  flowchart, 13
  government influences, 47–49
  large-scale influences, 43–52
  leader's role, 57–59
  Lean model, 74
  national cultures, 43–44
  organizational cultures, 46–47
  peers and friends, 51–52
  shift, 64
  summary, 59–60
current state, challenging, 58
customers, *see also* User/customer satisfaction
  behaviors, 36, 51
  data and analytics flowchart, 13
  focus on, 54, 58
  as judge, 164
  weakness awareness, 158
cyberspace and cyberworld, 1–2

D

DAS, *see* Defense Acquisition System
dashboards, 161–162
data collection, 206, 208–209, *see also* Information
decisions
  collaborative decision making, 13
  longer-term perceptions, 30
  risks and consequences, 35
  satisfaction, 28–29

Defense Acquisition System (DAS)
  capabilities and capability gaps, 85, 86
  needs and requirements satisfaction, 82, 86–88
  relationship management, 90–91
  website, 89
Defense Acquisition University (DAU), 100
demand signals, 92
demographic questions, 21–22
diagnostic elements
  active data sources, 134–135
  assessment, 9, 136
  causal inquiry, 138
  environmental scan, 133–134
  information summary, 135–136
  passive data sources, 135
  profiling the organization, 136–138
  situational analysis, 132
directive driven innovation strategy, 146, 147–148
dirigibles, 231
disciplined innovation approach, 202
disconnect, 153
discussion questions, *see end of each chapter*
disruptive innovation
  existing strategies, 143, 144–145
  innovation types, 20
  macroinnovations, 198, 201
diversity policies, 44–45
doctor examples, 131, 154
documentation, affinity for, 68–69
dominance criteria, 46
drawdowns, military, 228, 231, 233
driver's license renewal example, 20
dynamic capabilities, *see also* Capabilities; Learning
  basic concepts, 6–7, 63–66

cause and effect, 74–78
flowchart, 13
leadership, 66–68
Lean concepts, 72–74
process mapping, 68–72
summary, 78
dynamic vs. static, 38

E

educational institutions, 229
effect, estimation, 52–57
effective innovation, 4
efficiency and effectiveness, 83, 162
embedded contractors, 98–99
embedding, secure, 158
emotional attachment, element of satisfaction, 29
employee awareness, competency, 160
empowerment, 55, 158
ENOVALE™: How to Unlock Sustained Innovation Project Success, 183, 205, 217
ENOVALE™ approach, 11, 168
environment, *see* Culture and environment
environmental scan, 133–134
establish controls, 214
estimating effects, culture and environment, 52–57
ethnicity
    basic concepts, 5–6
    culture and environment, 44–46
evaluate and execute
    innovation readiness, 118
    organizational assessment, 104–112
    project selection, 192–193
evolutionary complexity, 154
evolutionary process, 12
execute, *see* Evaluate and execute

existence, organizational purpose, 150
existing strategies, 143–145
expectations
    belief formation, 35
    culture and environment, 54–55
    evaluation, 185–188
    and perceptions, 30–31
experiences, judgment, 34–36
explicit-to-tacit knowledge, 65–66

F

FAA, *see* Federal Aviation Administration
Facebook, 134
failure modes and effects analysis (FMEA), 183
family, 51
fast second strategies, 143, 144
FCB, *see* Functional Capability Boards
FCC, *see* Federal Communications Commission
fear, freedom from, 53
features, intangibles, 37
Federal Aviation Administration (FAA), 230
Federal Communications Commission (FCC), 230
federal community and government
    case study, 11–12
    catalyst for innovation, 2
    future developments, 227–234
    government civilians, 98–99
    government influences, 47–49
    summary, 234
Federal Knowledge Management Working Group (FKMWG)
    background, 21–23
    basic concepts, 2
    charter, 2
    ongoing research, 5

244 ■ *Index*

overview, 20–25
results summary, 23–25
femininity, 43
filtered knowledge, 33–34
financial results, as driver, 151
fiscal prudence, 3
fishbone diagram, 75–76
fixed-wing aircraft example, 8
flowcharts
    alignment and adaptation, 186, 212
    AREA, 218
    assessment process, 137
    diagnostic evaluations for innovation, 130
    establish controls, 214
    evaluate and execute, 193
    identifying/acting on needs (requirements), 26
    incremental innovation process, 13
    innovation evaluation and satisfaction with decision, 28
    needs, 170–171
    nominate and negotiate, 204
    nominate and normalize, 173
    objectify step, 175
    operationalization, 176, 208
    reasons and requirements, 205
    replacement cycle, 215–216, 218, 220, 222–224
    starting point, 194
    track and tie to performance, 212–213
    validation, 179, 183, 211
    verification, 211
FMEA, *see* Failure modes and effects analysis
F-22 Raptor aircraft, *see* Aircraft parts procurement case study
"free-for-all" mentality, 77
freelancing, 53
friends, *see* Peer pressure
Fuji company, 145
Functional Capability Boards (FCBs), 89
functional-user relationship, 177–178
function-related questions, 21–22
functions, compatibility with, 153–155
future developments, 227–234
Future Years Defense Program (FYDP), 90
FYDP, *see* Future Years Defense Program

G

generational differences, 199
generation of requirements, 93–97
generations X and Y, 233
globalization, 42, 64
government civilians, 98–99
government influences, 47–49, *see also* Federal community and government
"grid approach," 122–123

H

healthcare system, 231
helicopter program, 97
hidden dynamic capabilities, 66
Hispanic values, 45–46
history criteria, 46
Hofstede model, 43–44
"home-grown" approach, 145
home offices, 228
human impact analysis, 185

I

ICD, *see* Initial Capabilities Document

ideas, 17
illness example, 131
impact of innovation, 198–200
IMPACT test, 220–221
implementation
   basic concepts, 202–203
   incremental improvement, 203–213
   innovation strategies, 159–160
   N²OVATE™ methodology, 202–223
improvement, Lean model, 73
inclusion, 38
incremental improvement and innovation
   basic concepts, 12–13, 200
   as best "fit," 19
   N²OVATE™ methodology, 203–213
individualism/collectivism, 43
influencers, 42, *see also* Culture and environment
information
   active sources, 134–135
   collecting, 31–32
   importance of, 31–32
   passive sources, 135
   summary, organizational diagnostics, 135–136
Initial Capabilities Document (ICD), 87
initiating innovation, *see* Needs (requirements)
innovation
   combination of perspectives, 15–39
   creation, 36
   difficulty achieving, 4
   effective, 4
   incremental, 12–13, 19, 200, 203–213
   process management, 11–12
   recognition, 37
   strategies, 10
   success, 108–112, 141
   sustainability, 36
Innovation Processes and Solutions, LLC (IPS), 123, 129, 174
innovation readiness (IR)
   actions, 112
   align and adopt, 117
   basic concepts, 114–115
   evaluate and execute, 118
   flowchart, 13
   interpretation of results, 119, 122
   needs (requirements) and new ideas, 115
   nominate and normalize, 115–116
   objectify and operationalize, 116
   scoring, 118–119
   track and transfer performance, 117–118
   verify and validate, 116–117
innovation strategies
   ad hoc, 146, 147
   assembling the strategy, 158–159
   basic concepts, 10, 141–142
   compatibility, outcomes and functions, 153–155
   constructing, 155–159
   directive driven, 146, 147–148
   existing strategies, 143–145
   flowchart, 13
   implementation, 159–160
   lack of, 146, 148–149
   mission, vision, purpose statements, 150–152
   organizational scale, 142–145
   persistence, 159
   research & development emphasis, 146, 148
   science based, 146, 148
   success factors, 161–164
   summary, 164
   supplier based, 146, 147

SWOT analysis, 156–158
types, 145–149
innovative culture, 46
intangibles, *see* Performance
integrated approach, *see*
Incremental improvement
integration, *see* IMPACT test
intellectual capital, 56
Internal Revenue Service (IRS), 152
involvement, competency, 160
i-products
existing strategies, 144–145
innovation impact, 200
long-term alliances, 56
new application, 18
IR, *see* Innovation readiness
IRS, *see* Internal Revenue Service
Ishikawa diagram, 75–76

J

jet aircraft example, 34
job classification variable, 22–23
Joint Capabilities and Integration
Development System
(JCIDS)
capabilities and capability gaps, 86
Defense Acquisition System, 87
needs and requirements
satisfaction, 82, 88–89
relationship management, 90–91
requirements generation process, 93–97
Joint Capability Areas (JCA)
approach, 89
Joint Requirements Oversight
Council (JROC), 88
joint ventures, 54
judgment
basic concepts, 33
experiences, 34–36

influence, innovation evaluation, 16
knowledge, 33–34
performance, 36–37
"jumping in" approach, 103
justification, 95–96

K

Kawasaki company, 145
key contributor categories, 76–77
key performance indicators (KPIs),
see Measures and metrics;
Success factors
key performance parameters (KPP), 88
knowledge
influence, innovation evaluation, 16
judgment, 33–34
vs. position, 5, 24
knowledge age, 1
KPP, *see* Key performance
parameters

L

lack of performance, 204, 206
lack of strategy approach, 146, 148–149, 152
ladders of trust, 56, *see also* Trust
Lambaria, Anastacio ("AL"), 96
language, 71, 150–151, *see also*
Process mapping
large-scale influences, 43–52
latent dynamic capabilities, 66
leadership
challenges, 98–99
culture and environment, 57–59
dynamic capabilities, 66–68
by example, 57–58
mentoring, 95
mindset shift, 64

replacement step success, 221–222
requirements generation process, 98–99
strategy success elements, 158
succession, 95, 98
Lean concepts, 72–74
"leapfrog" innovations, 144, 201
learning, *see also* Dynamic capabilities; Skill sets; Training
　knowledge acquisition, 34
　"learn-as-you-go" endeavors, 188
　from mistakes, 159
lessons learned, *see* Learning
life cycles, new innovation, 18
Likert scales
　innovation strategy types, 146
　work environment, 112–113
limitations, 179, 181–182, 209
lip service, 152
longer-term perceptions, 30
long-term orientation, 44
loyalty, element of satisfaction, 29

M

macroinnovation
　impact of, 198
　N²OVATE™ methodology, 199, 200–201
　process for each, 194
　subdividing projects, 10
managed change, *see* IMPACT test
Marines, 88
masculinity/femininity, 43
material development decision (MDD), 87
MDD, *see* Material development decision
measures and metrics
　alignment, 185
　basic concepts, 10

culture, 46–47
nonexistent or poor, 19
objectives, 142
persistence, 159
success factors, 162–163
user/customer satisfaction, 27–28
medical office example, 154
Medicare, 231
mentoring and succession plan, 95
microinnovation
　impact of, 198
　N²OVATE™ methodology, 199–200, 202
　subdividing projects, 10
Microsoft Windows NT, 200
military
　budget submission, 94
　buildups and drawdowns, 228, 231, 233
　training academy, 229
millennial generation, 233
mission
　innovation strategies, 150–152
　needs and requirements satisfaction, 82–85
mistakes, learning from, 159
Mitsubishi company, 145
models, experiences as, 35
mood, 134

N

National Aeronautics and Space Administration (NASA), 151–152
national cultures, 43–44
Navy, 88
needs (requirements)
　basic concepts, 81–82
　budget, 89–90
　capabilities, 85–86
　complexities management, 90–91
　core competency, 82–85

criteria, 25
Defense Acquisition System, 86–88
execution process, 89–90
flowchart, 13
innovation process, 7–9
Joint Capabilities and Integration Development System, 88–89, 93–97
leadership challenges, 98–99
mission set, 82–85
new ideas, 115
perspectives, 25–27
planning and programming, 89–90
requirements, 82–85, 93–97
sequestration, 91–93
summary, 99–101
values assessment, 124–126
vs. wants, 81, 84, 96
newness, 18
nominate and normalize
  incremental improvement, 201–202
  innovation readiness, 115–116
  project selection, 170–174
noncombatant agencies, 89
normalize, *see* Nominate and normalize
no strategy approach, 146, 148–149, 152
N²OVATE™ methodology
  adaptability, 11, 155
  align and adapt, 117
  basic concepts, 168, 197–198
  capabilities and capability gaps, 85
  culture assessment, 112–114
  Defense Acquisition System, 87
  evaluate and execute, 118
  evaluating success, 108–112
  flowchart, 13
  impact of innovation, 198–200
  implementation, 202–223
  incremental improvement, 203–213
  innovation readiness, 114–124
  introduction, 3
  Lean concepts, 72–73
  macroinnovation, 199, 200–201
  microinnovation, 199–200, 202
  modifications, 193, 197, 203
  needs (requirements), 115
  nominate and normalize, 115–116
  objectify and operationalize, 116
  organizational assessment, 122–124
  organizational readiness, 104–108
  replacement, 214–223
  requirements, 8
  scoring, 118–119
  success, evaluating, 108–112
  summary, 223–225
  track and transfer performance, 117–118
  verify and validate, 116–117
Novell, 200

O

objectify and operationalize
  incremental improvement, 206, 208
  innovation readiness, 116
  project selection, 174–178
objectivity, 65
offices, traditional, 228
"old guard" mentality, 99
"1–n" list, 96–97
"one-size-fits-all approach, 10, 154, 213
ongoing research, 5
open communications, 48, *see also* Communications

operating limits, 209
operationalize, *see* Objectify and operationalize
Operations and Support (O&S), 86–87
opinions
  belief vs. experiences and knowledge, 77
  evaluation of innovation, 16, 17–20
opportunities, *see also* SWOT analysis
  innovation as, 38
  value-added, 18–19
organizational assessment
  actions, 112
  align and adopt, 117
  basic concepts, 9, 103–104
  culture, 46–47, 112–114
  diagnostics, 136
  evaluate, 104–112, 118
  execute, 118
  flowchart, 13
  innovation readiness, 114–122
  innovation success evaluation, 108–112
  needs (requirements), 115, 124–126
  new ideas, 115
  nominate and normalize, 115–116
  N²OVATE™ IR proficiency, 122–124
  objectify and operationalize, 116
  readiness for innovation, 104–108
  summary, 126–127
  track and transfer performance, 117–118
  values assessment, 124–126
  verify and validate, 116–117
organizational diagnostics
  active data sources, 134–135
  assessment, 136
  basic concepts, 129
  causal inquiry, 138
  diagnostic elements, 132–138
  environmental scan, 133–134
  flowchart, 13
  information summary, 135–136
  passive data sources, 135
  profiling the organization, 136–138
  reasons for, 131
  situational analysis, 132
  summary, 139
organizational existence purpose, 150
organizational scale, 142–145
Organization Science, 67
O&S, *see* Operations and Support
outcomes
  completion criteria, 164
  defining, 109
  evaluation, 160
  flowchart, 13
  impact of, 180, 182
  innovation strategy compatibility, 153–155
  performance measures, 211
  success factor, 162
  validation and verification, 178–185

P

PAR, *see* Presidential Aircraft Recapitalization program
paradigm shift, 64
paradox, 15
parents, *see* Family
partnering, 55
parts procurement case study, 168
passengers example, 34
passive data sources, 135
patterns, 134

peacekeeping efforts, 230
peer pressure, 51–52
perceptions, *see also* Perspectives
   basic concepts, 4–5
   evaluation, 185–188
   and expectations, 30–31
performance, *see also* IMPACT test; Track and transfer performance
   judgment, 36–37
   tangible aspect, 19
persistence, 158, 159
personal values, element of satisfaction, 29
personnel cuts, 3, *see also* Workforce reduction
perspectives, coalescence of
   basic concepts, 15–16
   choices, amalgamation of, 30–31
   coalescence/combination, 16–20
   experiences, 34–36
   inclusion, 38
   information importance, 31–32
   judgment, 33–37
   knowledge, 33–34
   needs/requirements satisfaction, 25–27
   performance, 36–37
   research study, 20–25
   satisfaction, 25–27
   summary, 38–39
   user/customer satisfaction, 27–30
physician examples, 131, 154
picture, big, 59
Pines, 200
Pixar company, 68
Planning, Programming, Budget, and Execution (PPBE)
   capabilities and capability gaps, 85
   needs and requirements satisfaction, 89–90
   relationship management, 90–91

police officer example, 132
policies
   complexity example, 154
   directive driven strategy, 146, 147–148
   diversity policies, 44–45
   organizational scale, 142–143
   political systems influence, 48
   strategy success elements, 158
political system influences, 47–49
populace receptiveness, 48
position vs. knowledge/skill sets, 5, 24
power distance, 43
PPBE, *see* Planning, Programming, Budget, and Execution
preferences, element of satisfaction, 29
Presidential Aircraft Recapitalization (PAR) program, 8
presidential rotary program, 97
pressure to conform, 51
price, 125
pricing, element of satisfaction, 29
priorities, changing, 188
private-sector business practices
   exploration of, 3
   public-sector contrast, 3
private vs. public sector, 3
procedures
   dynamic capabilities, 6–7
   political systems influence, 48
processes
   dynamic capabilities, 6–7
   management, 11–12
   structure, 155
process mapping
   benefits, 70–72
   importance, 68–70
procurement example, 217
product structure, 155
professional association variable, 22–23

profiling the organization, 136–138
profitability (cost effectiveness), 35
proven success rate, competency, 160
public domain, alliances in, 55–57
public vs. private sector, 3, 190–192
purchases, choices, 31
purpose
  Lean model, 73
  organizational existence, 150
  statements, 150–152

Q

questions, *see also* Innovation readiness
  cause and effect, 74–78
  leadership careers, 227
  SWOT analysis, 157
questions for discussion, *see end of each chapter*

R

radar chart, 119, 1456
radical complexity, 154
R&D, *see* Research & development emphasis
readiness for innovation, *see* Organizational assessment
real-time analytics, 44
real-time information, 32, *see also* Information
reasons, *see* Requirements
receptiveness to populace, 48
recognition, innovation, 37
regional affiliations, 45
rejection, choices, 31
replacement
  N²OVATE™ methodology, 214–223
  positive outcome, 20

repressive regimes, 48
request for proposal (RFP), 84
requirements, *see also* Needs (requirements)
  basic concepts, 82–85
  detail of process, 177
  generation process, 93–97
  leadership challenges, 98–99
research & development (R&D) emphasis, 146, 148
research studies
  ongoing, 5
  perspectives, 20–25
resources
  competency, 160
  pooling, 56
respect, 54
return on investments (ROI)
  culture and environment, 54
  private-sector success, 3
  private vs. federal sector, 190–192
reviews, 160, 192, 216
RFP, *see* Request for proposal
risk
  avoidance, 162
  SREM diagram, 183–185
  taking, 53
rollout phase, 222
"rose-colored glasses," 34
routines
  dynamic capabilities, 6–7
  emphasis on, 65
rules of engagement, *see also* Standard operating procedures
  changes, 2
  defined and accepted, 77
  sequestration, 92

S

Samsung company, 56

252 ■ Index

satisfaction, *see also* Needs (requirements)
  elements of, 29
  influence, innovation evaluation, 16
  users (customers), 27–30
science-based innovation strategy, 146, 148
scorecards, 161–162
scoring, *see also* Surveys
  culture assessment, 47
  innovation readiness, 118–119
scripted processes, 233
secure embedding, 158
selection, innovation projects
  adaptation and alignment, 185–190
  basic concepts, 167–169
  change acceptance, 188, 190
  execute and evaluate, 192–193
  normalize and nominate, 170–174
  objectify and operationalize, 174–178
  summary, 193–194
  tabulate and track performance, 190–192
  validate and verify, 178–185
self-serving attitude, 98
sequestration
  defined, 63
  as driving force, 3
  FKMWG, 2
  needs and requirements satisfaction, 91–93
  realities, 230
service structure, 155
Shook's Lean model, 73–74
short-term events, 134
short-term orientation, 44
short-term perception, satisfaction, 29
significance criteria, 46

"silo" mentality, 92
situational analysis, 131, 132
Six Sigma, 68, 74
skill sets, *see also* Learning
  dynamic capabilities, 7, 65–66
  vs. position, 5, 24
SMART criteria, 174–176
SMPA, *see* Success Modes and Performance Analysis tool
SNIFF test, 214–215
social "cool factor," 144
Social Security, 231
SOP, *see* Standard operating procedures
SOW, *see* Statement of work
Spirit company, 145
sports drink example, 51
SREM (success, risk evaluation matrix) diagram, 183–185
standard operating procedures (SOPs), *see also* Rules of engagement
  basic concepts, 83
  dynamic capabilities, 7
statement of work (SOW), 85
"state of mind," 134
static vs. dynamic, 38
status quo, choices, 31
strategic structure, 155
strategies, *see also* Innovation strategies
  measuring defined requirements, 96
  success elements, 158
strengths, scoring, 118, *see also* SWOT analysis
"structural holes," 55
structure categories, 155
substitution complexity, 154
success factors
  basic concepts, 10
  communication, 67–68
  competencies, 160

Index ■ 253

    innovation strategies, 161–164
    shared characteristics, 37
    SREM diagram, 183–185
succession plan, 95
Success Modes and Performance
       Analysis (SMPA) tool,
       206–207
summaries
    culture and environment, 59–60
    dynamic capabilities, 78
    federal government, 234
    innovation strategies, 164
    needs and requirements
       satisfaction, 99–101
    N²OVATE™ methodology,
       223–225
    organizational assessment,
       126–127
    organizational diagnostics, 139
    perspectives, 38–39
    project selection, 193–194
supplier based innovation strategy,
       146, 147
supportive culture, 46, 95
surveys, *see also* Scoring
    acceptance of change, 189–190
    expectations and perceptions,
       186–188
    FKMWG study, 5
    innovation comprehension,
       171–173
    missed opportunities, 36
    normalize and nominate step,
       171–173
    passive data, 135
sustainability
    assessment, 26, 169
    culture and environment, 53–54
    innovation, 36
SWOT analysis (strengths,
       weaknesses, opportunities,
       threats), 155–158
systems structure, 155

T

tabulate, project selection, 190–192
tacit-to-explicit knowledge, 65–66
tactical innovation, 167
tangibles, *see* Performance
taxes, 48
temperature example, 134
threats, *see* SWOT analysis
3M company, 68, 200
time, 54, *see also* IMPACT test
"time in the cockpit," 24
timeliness, 83
TOA, *see* Total obligation authority
tools, 54
top-down approach, 53
total obligation authority (TOA), 90
Toyota, 72
track and transfer performance
    incremental improvement, 213
    innovation readiness, 117–118
    project selection, 190–192
traditional offices, 228
traffic accident example, 132
training, *see also* Learning
    culture and environment, 54
    dynamic capabilities, 54
    educational institutions, 229
transfer performance, *see* Track and
       transfer performance
tribal affiliations, 45
TRICARE, 231–232
trust, 49, 54
trust ladders, 56

U

UAV, *see* Unstaffed aerial vehicles
uncertainty avoidance, 44
uncontrollable traits, 44
underperformance, 204, 206
unplanned events, 96
unplanned risks, 35

unquestionable direction, 233
unsatisfied needs/requirements, 7–9
unspoken rule, 99
unstaffed aerial vehicles (UAVs), 231
users/customers, *see also* Customers
   competency, 160
   functional-user relationship, 177–178
   innovation must please, 24–25
   as judge, 164
   perspectives, 27–30

V

validate, *see* Verify and validate
value added (competitiveness), 35
values
   ethnicities, 45–46
   multiple meanings, 125
variation, strategy type, 149
VC-25 example, 8–9
verbal cues, 133
verify and validate
   incremental improvement, 210–211
   innovation readiness, 116–117
   project selection, 178–185
Veterans Affairs (VA) healthcare systems, 231
viability
   assessment, 27, 169

political systems influence, 48
vision, innovation strategies, 150–152
visual assessment, 133
"voice of the operator," 24
Vought company, 145

W

weaknesses, scoring, 118, *see also* SWOT analysis
websites
   active data sources, 134
   Defense Acquisition Portal, 88
   Defense Acquisition System, 89
   Defense Acquisition University, 100
"what if" analysis, 131, 138
white space, 54, *see also* Blue oceans
Why questions, 74, *see also* Cause and effect
Windows NT (Microsoft), 200
work environment, *see* Culture and environment
workforce reduction, 229, 232, *see also* Personnel cuts

Y

yearly reviews, 160